# They LOVE and KILL

BY VITUS B. DRÖSCHER

*Translated from the German by
Jan van Heurck*

# They LOVE
# and KILL

*Sex, Sympathy and Aggression*
*in Courtship and Mating*

**Vitus B. Dröscher**

*A Sunrise Book*

**E. P. Dutton & Co., Inc.  ·  New York**

*To my wife*

Drawings by Helmut Skaruppe

First published in the U.S.A. 1976 by E. P. Dutton & Co., Inc.

*Sie töten und sie lieben sich* copyright © 1974 by Hoffmann
und Campe Verlag
English translation copyright © 1976 by E. P. Dutton & Co., Inc.,
and W. H. Allen & Co. Ltd.

First Edition

10 9 8 7 6 5 4 3 2 1

Library of Congress Cataloging in Publication Data

Dröscher, Vitus B
  They love and kill.

  Translation of Sie töten und sie lieben sich.
  "A Sunrise book."
  Includes bibliographical references and index.
  1. Social behavior in animals. I.Title.
QL775.D7213          591.5          76–5806

ISBN: 0–87690–215–8
Designed by The Etheredges

Sources of photographs in this book are as follows (names in parentheses denote photographers of pictures furnished by agencies or publishers):

℃ Anthony-Verlag, Starnberg am See°: Pages 97 (R. H. Berger); 283, below (Sobottka); 295, above (Hoppé); 321, above (Grenzemann); color photos following page 108: first photo (Schmidecker), fourth (Schlattau) and seventh (Gensetter); color photos following page 236: second photo (Mahr), fourth (Soyka) and eighth (S. Thamm). ℃ Vitus B. Dröscher, Hamburg/Amerika-Dienst: Pages 33; 249, above. ℃ Kai Greiser, Hamburg: Page 7. ℃ Dr. Hans Jesse, Cologne: Pages 67 (both); 87; 129; 151; 233; 321, below. ℃ Greta Robok, Hamburg: Page 199, above. ℃ Walter Sittig, Hannoversch Münden: Pages 21; 55; 165; 249, below; 283, above; 293; 295, below; 307 (both). ℃ Suddeutscher Verlag, Munich: Pages 187, above; 199, below; 319 (H. Thieme). ℃ Carl-Albrecht von Treuenfels, Frankfurt am Main: Page 115. ℃ Zentral Film-Austalt, Düsseldorf: color photos following page 108: fifth photo (D. Baglin); sixth (G. Heilmann) and eighth (P. Fera); color photos following page 236: third photo (Dr. Sauer), fifth (H. Reinhard), sixth (Leidmann) and seventh (Lummer). ℃ Günter Ziesler, Munich: Pages 43; 116; 127 (both); 187, below; second and third color photos following page 108; first color photo following page 236.

°All cities in Federal Republic of Germany.

# Contents

riage – The female chooses a mate in an arena of four hundred bridegrooms – A pair bond that lasts only two seconds – Beauty: dead end of evolution? – Birds of paradise and bower birds – Artificial adornment and the return to simplicity and monogamy

## 15. Courtship at Court 245
### *The Turkey System*

The turkey dictatorship prevents monogamy – Childhood battles determine the social and sexual status of adults – Sworn brotherhoods – Swift and violent courtship – A plentiful food supply favors polygamy – Landscape and climate help to shape social and sexual mores

## 16. The Unhappy Life of a Pasha 253
### *The Harem*

Rulers of harems steal their wives – Feeble pashas are deposed – From the harem to the sexual commune – Females as the male's private property – The pasha and his deputy – Males that are incapable of holding onto a female – Factors that shape a society: heredity, environment, and education

## PART SEVEN: FORMS OF MARRIAGE

## 17. Marriage Partners Who Do Not Know Each Other 269
### *"Local Marriage"*

Some animals become aggressive and ready to mate when they are in a certain place – The inhibition of aggression toward a sexual partner is the basis of "local marriage" – The stork is married to the nest – The ability to distinguish between members of one's own species is one precondition of a personal bond – Monogamy develops when a personal bond is linked to the tie to a place

## 18. Only Aggressive Mates Stay Together 278
### *Seasonal Mating*

The care of young binds a couple together – Offspring are not the only reason for a marriage – Pair formation that lasts a few minutes or for several months – Fidelity to a place unites the mates of the previous year – The more peaceable an animal is, the more indifferent it is to its mate – What is fidelity? – From seasonal to permanent marriage

# Introduction

*Sympathy, the Antithesis of Aggression*

Occasionally a scientist may "moonlight," taking time off from his usual duties to perform some simple experiment. Scientists engaged in apparently frivolous or irrelevant research have often made unexpected discoveries that rendered them world-famous almost overnight. This was the case with Frank A. Beach, a professor at the University of California Zoological Institute.[1]

In 1970, Professor Beach had been listening to his wife. Almost every day Mrs. Beach used to take her female beagle Jacky for a long walk in Berkeley's Tilden Park. In the evening she would tell her husband how Jacky treated male dogs she met in the park. Sometimes she snubbed them completely. At other times, she would play the tyrant and leave her fiery suitors dangling until she decided that they had suffered enough. Then she would reward them by consenting to play with them.

Professor Beach assumed that his wife, an ardent dog-lover, had read things into her dog's behavior that were not there. After all, he thought, a dog is only an animal, and to a female animal, one male is as good as another. The professor believed that the sexual relations of animals were indiscriminate; thus unlike human beings, dogs could not feel personally attracted to some dogs and reject the attentions of others. At most, he thought, a dog might experience momentary whims, a fleeting inclination or disinclination to play or mate with a particular partner. Professor Beach wanted to offer his wife scientific proof that dogs could not behave in the manner she had described. He began a series of experiments that eventually proved his wife's claims correct. The experiments also proved a great deal more.

Beach experimented with five male and five female beagles that he raised from puppyhood in a large enclosure. Keeping the dogs under close observation, he noted that all the males showed a desire to make friends with all the females. Thus the behavior of the males bore out the professor's theory concerning the indiscriminate nature of canine mating habits. However, the females were a different story.

Only one male, Alf, seemed consistently attractive to all the females, Anita, Bonny, Cilly, Dora, and Elma. He was the cock of the walk, the "playboy" of the group. On the other hand, only Anita and Elma liked Bonzo; the other three females rejected him. Caesar appealed only to Anita and Cilly. Dora, and usually Elma, were friendly with David. But notwithstanding his zeal, the fifth male, Enno, could not master the art of dealing wtih females and was rejected by all five.

Each of the females demonstrated affection for one, two, or three of the males. Their taste in males varied; but none of the females changed her feelings with the passage of time. Professor Beach's experiments established that after several years, the responses of each female to each male remained constant.

Professor Beach made his most surprising discovery when the young bitches first went into heat. As every dog-owner knows, male dogs always flock around a female in heat. However, the female is very choosy about how she distributes her favors. Professor Beach was surprised to see which males the female beagles selected as mates.

In the past, Cilly had enjoyed close friendship with Alf and

Caesar and had rejected the attentions of the other males. But when it came time to choose a sexual partner, she banished her two friends and mated with Enno and David.

In the past, Anita had shown affection for Alf, Bonzo, and Caesar. When she went into heat, she suddenly began to snub Alf and Bonzo and mated with Caesar, as well as with Enno, who had been the "loser" of the group.

When the bitches were no longer in heat, they reverted to their former behavior, being friendly with their old friends and rejecting their sexual partners with snarls and bites.

Thus female dogs clearly distinguish between friendly relationships based on personal sympathy or affection and relationships based on sexual attraction. They place sexuality and affection in different categories.

To be sure, in some cases the female beagle chose a playmate as her sexual partner. Apparently certain traits recommend a male dog as a playmate or a lover, and occasionally one male may fulfill the requirements of both roles. However, as yet we have no idea what these traits are.

Professor Beach's experiments revealed an important fact about the social behavior of animals. Two profoundly different forces can bind two animals together, the sexual bonding instinct and the social bonding instinct. The latter is based on a feeling of sympathy, or mutual affection and liking. Sexuality unites male and female for only as long as it takes them to mate, or at most for a single mating season. The social bond may endure for years and even for a lifetime.

Scientists investigating animals other than dogs have confirmed this momentous discovery: There exists a bonding instinct based on a feeling of mutual sympathy.

In 1963, in his book *On Aggression,* Konrad Lorenz [2] demonstrated the instinctual nature of aggression. His research in this and other areas earned him the Nobel Prize. In his determination to prove his point, Lorenz may have neglected to stress the influence of education and environment on aggressive behavior, and this omission has led some people to misinterpret his findings. But in any case, Lorenz did succeed in establishing that aggression is an instinct.

Some people mistakenly view the sexual instinct as the antithesis of aggression. In reality, the antithesis of aggression is the social bonding instinct, based on the feeling of sympathy or affinity between two individuals. Scientists have only recently discovered the existence of this instinct.

The social bonding instinct brings male and female, man and woman together and forges a bond transcending the bond created by sexual needs. Until recently, we knew even less about this instinct than we do about sexuality and aggression. In this book I attempt for the first time to examine the relationship between the three "enemies of reason": aggression, sexuality, and the sympathy bonding drive. I also discuss the relationship between these instincts and reason, will, the sense of shame, and our human responsibility for our acts.

The discovery of the social bonding instinct revolutionizes our time-honored views of love and marriage. By understanding the role this instinct plays in our lives, we may learn to understand how men and women can live together in greater harmony.

Some people may question the statement that the understanding of our instincts can help our society to function more harmoniously. Contemporary man is inclined to ignore the natural history, as well as the political history, of mankind. He tends to believe that human reason is all-powerful and that instinctual behavior patterns inherited from the past have little influence on modern marriage or, for that matter, on modern society as a whole.

To be sure, human reason is a powerful creative force. However, it is disastrous to ignore the destructive potential of man's intelligence. Like everything else, the workings of the human mind are subject to natural law. Ultimately, we are not rational creatures. Our intelligence may give us a certain amount of control over the forces of the unconscious; but up until now, we have failed miserably in our attempts to create a rational society.

We can control our instincts only if we learn to understand them. This book explores one such instinct, the social bonding instinct, as it functions in sexual relations. In a future book I will discuss the bond between parents and offspring, as well as that between members of larger societies.

To understand the irrational forces that govern us, we must study the animal kingdom, where these forces first evolved.

Reproductive behavior represents a major element in the natural history of social behavior. As life evolved, this behavior became increasingly complex. The history of reproductive behavior begins with sexual reproduction. Later developments include the "invention" of the male, cannibalistic mating, rape as the "normal" form of copulation, the birth of the sympathy bonding instinct, the evolution of various mating and marriage patterns, and finally, the development of analogous behavior patterns in man.

# PART ONE

# Nature Experiments with
# Male and Female

# CHAPTER 1

# Virgins Bear Young

*Primitive Forms of Reproduction*

There exist races of females which survive almost totally without the help of males. In many ways these females recall the legendary Amazons of the Homeric age, who were so strong and skilled in warfare that only the greatest heroes of antiquity—men like Hercules and Achilles—could hold their own against them.

During most of the year, the Amazons murdered all the men they met. However, once a year these man-haters showed the men of neighboring kingdoms how charming they could be when they wanted to. They could not get along without men altogether, for men were necessary to the conception of children. As soon as their children were born, the Amazons killed all the boys, and raised the girls to wage war against the whole race of men.

The legend of the Amazons may have been partly based on historical truth; or perhaps the legend merely reflects the age-old prejudices of men against women. In any case, in the world of fish, Amazons are a living reality. Of all the mythical monsters and biological chimeras conceived by the classical Greek imagination, there are few that do not really exist somewhere in the animal kingdom.

Early Hellenistic times were an age of transition, when men were outgrowing their dependency on myth and struggling to achieve a more scientific understanding of the world around them. Pondering the origin of the world, the nature of living creatures, and the relationship between man and woman, human beings dreamed up phenomena remarkably similar to creatures already devised in the infinite experiments of nature itself.

The living counterpart of the Amazons is a fish called the Amazon molly, which is closely related to the guppy. Approximately the size of a man's finger, the Amazon molly is native to Central America. The fish closely resembles the legendary Amazons except in one respect: The female does not kill her male offspring, for she never gives birth to males.

Amazon mollies are a race of females which give birth only to females. Of course, even they find males necessary to the conception of young; thus they "borrow" the males of other species closely related to their own. Once a year, overcoming their detestation of the opposite sex, the Amazon mollies pay a call on the males of their close relatives. In the rivers and coastal lagoons of southern Texas, the Amazons mate with male fish of the species known as *Poecilia latipinna*. In the fresh and salt water bodies of northeastern Mexico, they choose male black mollies.

The sperm of the male fish does not really fertilize the eggs of the Amazon mollies. The sperm penetrates the egg cells but only serves to activate the process of cell division, that process by which a single cell grows into the billions of cells that compose a living being. Once it has initiated this process, the sperm degenerates without fusing with the egg nucleus. Thus none of the hereditary traits contained in the sperm are inherited by the young. In most other species, the young inherit traits from both parents, but in this case all the male chromosomes are "murdered" in the female cell.

Thus although she does not remain a virgin, the Amazon

molly reproduces by means of unfertilized eggs. She reproduces not by parthenogenesis, but by a special kind of parthenogenesis known as gynogenesis. This strange conception produces a strange brood of young. All the young fish inherit exactly the same hereditary traits. Each sister resembles all the others down to the last jot and tittle, and each sister is a smaller replica of the mother. Amazon mollies exactly reproduce themselves. Their young inherit nothing from the father. Down through the millennia, the traits of each generation remain identical.

Our lives would seem a nightmare if human beings produced exact replicas of themselves like the Amazon molly. In *Brave New World,* his terrifying vision of the future, Aldous Huxley describes a class of human beings who have all been reduced to a single genetic norm. In a futuristic society, incubators are used to grow "Bokanowsky groups," a genetically and socially inferior caste of human beings developed to perform monotonous assembly-line work in factories. These completely identical creatures are bred to serve completely identical machines.

In various nations, biological institutes are performing experiments to determine whether parthenogenesis can be activated by artificial means. I will discuss these experiments later. For now, let us try to imagine what our lives would be like if human beings reproduced like Amazon mollies.

For one thing, the human race would consist only of women. Without human males, women would be forced to mate with chimpanzees or gorillas and would continue to reproduce replicas of themselves down through the generations. Relatives might live together in communities where every woman would look exactly like every other. Again and again we would encounter the same face and figure, the same mannerisms and speech habits, the same coughs and giggles, the same opinions and prejudices, the same tendency to temper tantrums, the same anxieties and the same tastes in fashion. Life would be a nightmare.

In our present world, before major elections we see the same face staring at us from a thousand billboards. If we reproduced like the Amazon molly, we would see the same faces all the time. We would exist in a world of types in which each type was reproduced ad nauseam. There would be no more individuality.

Identical twins often experience such a strong feeling of identity that each twin loses some of his individuality. Once when I asked two identical twins what they were doing, they answered in chorus, "We're looking for our shoe." Only one twin had lost a shoe, but both of them behaved as if they were the two halves of a single being.

In our population of human Amazon mollies, instead of identical twins there would be thousands of people who all related to each other as if they were identical twins. All these identical people would feel even less like individuals than identical twins do. Perhaps they would no longer even know that they were individuals. They might function selflessly like the denizens of insect societies, which are also composed of genetic stereotypes. Moreover, they would make perfect citizens for totalitarian states.

Some people might enjoy living in such a stereotyped society: demagogues, militarists, advertising agents, public-opinion pollsters, economic planning committees, and anyone trying to organize a group to get something done. However, passport officials and police hunting criminals would have a terrible time telling people apart!

An elephant courtship begins with a battle. The male and female elephants wrestle each other with their trunks. A first they engage in a violent struggle, then gradually begin to caress each other. Finally the bull and the cow stand head to head and gaze into each other's eyes. Since elephants have lashes one to two inches long, it looks as if they were winking slyly at each other. Suddenly the cow whirls around and crouches down, inviting the male to mount and mate with her. The honeymoon may last for several days. Afterwards the two elephants go their separate ways. The cow returns to the herd of females and young elephants, and the bull rejoins the other males. An elephant prefers the company of its own sex to the companionship of a monogamous marriage.

I have purposely drawn an exaggerated picture in order to show what the world was like before the "creation of the male."

Nature "invented" the male relatively late in the history of evolution. It is the female that bears young. Without the female, there can be no offspring. However, as I will shortly explain, the male can easily be dispensed with. In the beginning, our planet must have been populated solely by females. The invention of the male introduced certain improvements into the reproductive process, but at the same time created many problems.

Thus, much as the idea may offend male pride, Eve was not really created from one of Adam's ribs. Instead, it was the other way around. Moreover, the invention of the male, the luxury product of genetics, created many grave problems. In fact, we might say that the existence of the male, rather than the eating of the forbidden fruit, was what brought about the loss of Paradise.

The life of the Amazon molly shows us what a world without males is like. These fish possess no individuality. However, the fact that one Amazon molly looks exactly like another is far from tragic in itself. The life of these fish seems like a nightmare only if we imagine what human life would be like if we reproduced in the same way.

In the course of evolution, most species have found it advantageous to develop males. Why are there no males among the Amazon mollies? Does this species represent a primitive life form which has not yet evolved to the point of developing a male sex? The Amazon molly belongs to the fish family of livebearers. Males exist among all its close relatives. However, all these males are cannibals. When they see a female giving birth to young, they swim under the female's belly and gobble up the baby fish one after another the moment they come into the world.

To prevent the males from devouring their young, the females of the live-bearer family look for a hiding place before they give birth. The larger the fish population, the more difficult it is for the mother fish to find a safe hiding place. In other words, the greater the number of fish inhabiting a given region, the greater the number of young that will be devoured immediately after birth. The cannibalism of the males serves as a barbarous form of birth control and prevents overpopulation.

Amazon mollies do not have to worry that the males of their species might devour their young. They avoid contact with all males, briefly leaving their territory to mate with the males of closely related species, and then returning home at once. They ram and beat with their fins any male that dares to follow them.

The Amazon molly does not represent a primitive life form which has failed to develop sophisticated means of reproduction. Instead, it has achieved a special adaptation to prevent the young from being devoured by cannibalistic males. In any case, this species shows how an animal society can subsist without males.

Some populations of fish which are the ancestors of our common goldfish consist solely of females. These female societies have been found in the regions around the Ural and Caucasus Mountains, near Moscow, in Rumania, and in Brandenburg, Germany. No doubt they exist in other areas as well.

Probably Amazon mollies are descended from species in which there were males as well as females. To find other species without males, we must go back hundreds of millions of years in evolutionary history. These species represent very primitive forms of life. However, I should make one thing clear at the outset: The evolution of reproductive behavior does not constitute a steady progression from elementary to "higher" forms of life. To be sure, to some degree we are justified in speaking of evolutionary "advancement." After all, certain innovations, such as the invention of sexuality and of the male, have proved indispensable in the development of later species. On the other hand, the history of evolution is full of apparent "setbacks," of reversions to what a human being would regard as less advanced traits. For example, the greylag goose is monogamous and mates for life, and many of its sexual problems, like homosexuality, prostitution, and infidelity, recall similar problems of human beings. However, the more "advanced" chimpanzee, man's closest relative in the animal kingdom, practices free love rather than monogamy, and in this respect resembles human beings

less closely than does the greylag goose. (I must caution ad-
vocates of free love not to conclude that this form of sexual behavior is necessarily higher in the scale of evolution than monogamy.)

Thus evolution does not obey the dictates of human morality. That is, human beings would not necessarily consider the
sexual behavior of the more advanced animals morally superior
to that of the lower animals.

Many factors influence sexual behavior: the aggressiveness
of the two partners; their ability to soothe each other's aggression, inspire trust, and suppress each other's desire to flee; the
degree of their sexual harmony; and the strength of the sympathy bond between them. The environment also plays a crucial
role, for the food supply, the abundance of enemies, the climate
and the landscape all favor a particular form of relationship.
Finally, animals themselves can teach each other appropriate
forms of sexual behavior. For example, Japanese macaques can
adapt to either a patriarchal or matriarchal society.

The multiplicity of factors that influence living creatures
makes it difficult for us to analyze the behavior patterns of animals and to judge the role analogous patterns play in human
life. Thus readers of this book should keep in mind the com-
plexity of the forces involved in the evolution, both past and
present, of animal species.

In the beginning, males and sexuality did not exist. The
first living organisms reproduced asexually. To understand this
fact, we must first understand the nature of reproduction.

We may define reproduction as the consequence of the bio-
chemical property of certain large protein molecules to produce
exact replicas of themselves. All life as we know it—the millions
of individual plant and animal species and the differentiation
into male and female—is a manifestation of the effort of protein
molecules to reproduce themselves in the most efficient possible
way.

When the history of life on our planet began, the tendency
of nature was to duplicate all existing forms as exactly as possible. Billions of years ago, when the earth was populated by
one-celled organisms, there were no males or females. All these

tiny microscopic creatures were neuter, and each reproduced by dividing into two beings identical to itself.

Reproduction by fission offers organisms two enviable advantages: eternal youth and a kind of immortality. If an amoeba traced its ancestry back hundreds of millions of years, it would not encounter a single corpse. When we look at an amoeba under a microscope, we are seeing a creature that may have been born only one hour ago, when its parent cell split into two halves. The parent cell was born yesterday, when the grandparent split in two; in turn, the grandparent cell was born two days ago. Thus we could trace the family tree of a single amoeba, going back billions of generations, until we finally arrived at the moment when the first one-celled organisms came into being.

No direct forebear of any amoeba existing today has ever died. If an amoeba dies, it is unable to divide and thus produces no descendants. Poisonous substances, starvation, unfavorable climatic conditions, and enemies which feed on amoebas never claim the lives of parent amoebas, but only of their twins. However, an amoeba is immortal only in the sense that it has survived untold aeons. We can never predict that it will be immortal in the future. The only thing we can safely predict about its future is that it will not die of old age. If it does not starve to death and is not eaten by a predator, poisoned, or killed by a hostile climate, the tiny creature will remain eternally young; for a biological rule of thumb states that as long as an organism continues to grow, it does not age.

Many people have wondered why giant tortoises often live to be three hundred years old, whereas human beings rarely live even a third of that span. The answer is that giant tortoises cease growing only a few years before death. At the age of two hundred, they are still growing, albeit not as swiftly as they did at twenty. Their continued growth keeps them young. The same is true of crocodiles, which can live to be one hundred; and no doubt it was true of the dinosaurs as well. The giant sequoias of California, which continue to grow until shortly before they die, live to be as much as four thousand years old and may reach well over three hundred feet in height.

We can see the tiny amoeba only under a microscope. Yet despite the fact that it never becomes very large, it continues to grow for millions of years. As soon as its body reaches a critical size, it divides, grows, divides again, grows, and so forth; and it will go on growing until all life on our planet is extinct.

The simple structure of unicellular organisms enables them to divide and continually renew themselves, thus granting them eternal youth and immortality. Unfortunately, multicellular organisms cannot function in the same manner. The renunciation of youth and immortality is the price that highly developed life forms must pay for their sophistication. Every time we gain something, we lose something else. This law holds true even in the history of evolution. As life forms grew more complex, death by old age came into the world.

Before the development of multicellular organisms, nature revolutionized the reproductive behavior of the unicellular organism by creating sexuality without sexes.

Among the creatures that exhibit sexuality without sexes are the flagellates, unicellular organisms that propel themselves through liquid by means of vibrating hairlike processes or flagella. One of the flagellates is a pear-shaped organism that as a rule reproduces by fission. However, occasionally one of these tiny creatures rams the pointed end of its body into the rounded end of another of its species and creeps inside. Then the two organisms fuse to form a single individual. Human lovers sometimes cherish the illusion that the two of them can become completely one. Among some primitive organisms, this illusion is a reality.

Shortly after the two organisms have fused, the new individual reproduces by fission. Thus first, as in the books of magic, two become one. Then one becomes two again.

Do these pear-shaped organisms which fuse to reproduce represent "males" and "females"? What is a female and what is a male?

We can define a female as an egg-producing individual, and a male as a sperm-producing one. By this definition, organisms which fuse to form a single individual have no sex. However, we could regard the organism that infiltrates the second organism from the rear as a sperm cell, and its partner as an egg cell. In this sense, the two organisms do in fact constitute male and female. However, if the so-called "male" had encountered a different partner, it might just as easily have played the role of "female" to the other's "male." Thus the term "sexuality without sexes" implies that two organisms exhibit something akin to sexual behavior, although neither has a determinate sex.

Despite the fact that flagellates have no determinate sex, they do exhibit the rudiments of specialization, of the differentiation into male and female roles. Under the microscope scientists have detected organisms whose rounded rear parts are marked with a dark circle, a sort of "target" for a partner to aim at. Organisms marked with these rings show a tendency to play the female role, just as the organisms without markings tend to play the male role. Of course, a small "male" can treat a larger "male" as a female and slip inside it. Moreover, frequently a "female" permits another "female" to creep inside it. However, no small "female" ever penetrates a large "male." Thus one of the two partners may exchange roles, but not both.

Among the flagellates, a "father" and a "mother," or a masculine and a feminine organism, fuse to form a neuter being which, at least initially, reproduces by fission. But the organism formed by the fusion of two other organisms constitutes a completely new creature. It contains elements of both the father and the mother; it is both of them and yet neither of them. It is an individual in its own right.

This development represents a revolution in evolutionary history. The parent no longer serves as a sort of stencil used to duplicate thousands of completely identical offspring but rather acts as one of many brightly colored paints on the genetic palette.

The development of sexuality without sexes accelerated evolution. Henceforth life forms could change and develop without waiting for the accident of a beneficial mutation. By fusing, two individuals could create an individual with a new set of traits, and this individual in turn could help to create new combinations.

Sometimes unicellular organisms do not actually fuse to create new individuals. They may simply exchange "experience," as do some disease-breeding bacteria. As a rule, these unicellular bacteria reproduce by fission. However, sometimes two bacteria line up side by side like two ships, and by an act of "pure" or asexual sexuality, exchange genetic material through their cell walls. Afterwards the two organisms go their separate ways; but now they are new individuals whose chromosomes have been altered.

A bacterium that causes typhus may develop immunity to a particular antibiotic, and then, in a veritable orgy of group sex, pass on its immunity to others of its kind. As a result, the antibiotic will no longer effectively combat the disease. This form of "microbial sex" is advantageous to the bacteria themselves but constitutes a serious problem for physicians and pharmacologists.

The development of multicellular organisms rendered even more complex the problems of reproduction. Suddenly the division of a cell no longer represented an increase in the number of individuals. Instead, new cells all contributed to the growth of a single individual. Under these conditions, how could an organism produce offspring?

Probably it took nature millions of years to develop solutions to this problem. Logically, it seems impossible that multicellular life forms could ever have come into being. After all, which came first, the chicken or the egg? Multicellular organisms cannot exist unless they are capable of producing offspring. But multicellular organisms can develop a method of reproduction only if multicellular organisms exist. Consequently, such organisms ought properly not to exist at all.

Nevertheless, the improbable occurred; how, we do not know. Of the early species created in the great experimental laboratory of nature, a few have survived to the present day. These species give us some idea of what went on in the past. Remarkably enough, some of them bear a striking resemblance to the mythical monsters of Greek antiquity.

According to legend, a giant serpent with nine heads attached to nine long necks used to inhabit the marsh land near modern Argos. This creature was the Hydra. When some bold warrior cut off one of her heads, two new heads instantly sprouted from the wound and devoured the hero.

Such terrifying monsters do in fact exist. However, they are terrifying only to water fleas, for their tiny bodies are no more than three-quarters of an inch in length. They are freshwater polyps known as hydras.

But small as they are, these denizens of ponds and pools can do something that the mythical monster of the Greeks could not. If one of their tentacles is cut off, several new limbs will grow back in its place. Moreover, a complete new polyp will grow from the severed tentacle. If we cut a polyp into two hundred tiny pieces no larger than specks of dust, two hundred

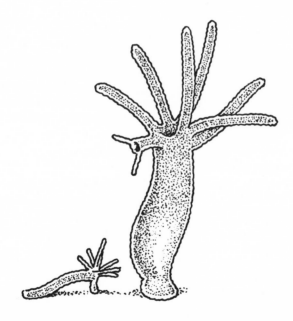

Reproduction by mutilation. The freshwater polyp is less than one inch long. When it loses one of its tentacles, several tentacles grow in place of the severed limb. Moreover, the severed tentacle grows into a new polyp.

hydras would be born from the remains. A hydra practices reproduction through mutilation.

If a fish "smacks his lips" a little when eating a hydra, sprinkling crumbs of his dinner into the water, he helps the polyp to produce many offspring.

However, as a rule the hydra produces its young by gemmation. A small budlike protuberance, or gemma, grows from the parent, becoming a baby polyp that finally separates from the mother's body. We could just as easily call it the father's body, for in this species males can also have children.

The only way to tell whether a hydra is a male or a female is to observe its behavior during a population explosion. If there is sufficient food and the weather is warm, the number of hydras in a pond may redouble each day. At this point, they are reproducing by gemmation. As the hydras exhale, carbon dioxide accumulates in the pond. When the gas reaches a critical concentration, it acts as a sexual stimulant, temporarily transforming the hitherto sexless hydras into males and females. The gas stimulates the polyps to develop swellings on the outer side of their stomach walls. Depending on whether the polyp was destined to be a male or a female, sperm or eggs form inside the swellings. As soon as the eggs and sperm are mature, they are expelled into the water. From then on, chance decrees whether or not the eggs will be fertilized.

Thus when overpopulation occurs, these sexless creatures suddenly become male and female. When there are so many of them that they need not fear extinction, they can afford the luxury of sexuality. When their numbers dwindle, the polyps become sexless again.

To be sure, under these conditions, we cannot attribute true sexuality to hydras. We can only say that under certain circumstances, they temporarily develop traits analogous to maleness and femaleness.

Only if the hydra population is very dense do the eggs and sperm have a chance of meeting so that fertilization can occur. In times of overpopulation, sexual reproduction offers one advantage over asexual methods. The delicate polyps and their gemmae die in cold weather. They could not survive a winter in

Europe, North America, or Siberia. But their fertilized eggs can survive the cold. Thus occasional sexual reproduction enables the polyps to leave tropical climes and populate cooler regions of the earth.

On the other hand, in other species, reproduction without the union of male and female cells may be more advantageous. In summer and spring, if they are to avoid extinction, plant lice must multiply so fast that they cannot afford to waste time looking for and wooing sexual partners and going to the bother of mating. Thus in these seasons, masses of females gather to suck the sap from the young shoots. Without ever having contact with a male, the females bear living female young.

Like unicellular organisms and the Amazon molly, female plant lice mass-produce their young, creating duplicates printed from a single pattern. All these virginal creatures produce exact replicas of themselves.

In the cool days of autumn, the female plant lice go into labor again. Their writhing bodies give birth to young males, which soon begin to mate with the females. This far from immaculate conception transforms the child-bearing virgin into an egg-laying female. As is the case with hydras, only the eggs of the lice survive the winter; all the existing plant lice die.

When a species alternates its reproductive method, producing first a dioecious generation and then a generation which reproduces by parthenogenesis, zoologists say that it is heterogonous.

It should be emphasized that the original purpose of sexual reproduction was not, as is often claimed, the avoidance of inbreeding. Instead, the development of sexuality enabled organisms to winter in cold climates. Nature had no need to create a special device to inhibit breeding, for creatures which develop degenerate traits as a result of inbreeding never survive for very long.

In the tropics, it is unnecessary for insects to produce eggs capable of surviving the winter, and males are not needed to guarantee survival. Thus in tropical regions, various species exist almost independently of males.

For example, certain species of stick and leaf insects, sawflies, gall wasps, and ichneumon flies, are races of Amazons in which the ratio of females to males is a thousand to one. In some of these species, at times only females are born from the unfertilized eggs; at other times only males. Thus periods when there is a large surplus of females alternate with periods when there is a surplus of males, as if the insects could not make up their minds which condition was better.

Bees and ants do not live in such confusion. In bee and ant societies, females develop from fertilized eggs; unfertilized eggs produce males. A signalling system regulates the production of the different castes. Thus the society produces no more of any caste than it needs.

A species of stick insect native to Australia, *Anchiole maculata,* reproduces by parthenogenesis for several generations. Then, for some unfathomable reason, the female offspring become infertile, and males must be found to save the population from extinction.

The females of certain species of leaf insects and bagworm moths mate with males when these are available. However, when males are in short supply, the females have no intention of becoming wallflowers. They simply produce their offspring by parthenogenesis, giving birth exclusively to males, which then mate with the females. These female insects take a do-it-yourself approach to the problems of reproduction: They produce their own males to marry!

In citing these examples of bizarre sexual behavior, I wish to demonstrate that the creation is a gigantic experimental laboratory where the relations between the sexes have undergone innumerable metamorphoses. Instinctual behavior patterns, modified by environmental influences, have determined the form of reproduction best suited to each species.

My discussion of the behavior of leaf insects and bagworm moths may have confused the reader. How can a female organism produce male offspring by parthenogenesis? After all, the case of the Amazon molly showed us that parthenogenesis creates exact replicas of the mother, and thus that all the offspring must be females. Studies of the bagworm moth have provided the answer to this question. The egg of the female moth contains two nuclei instead of one. During meiosis, the two nuclei fuse to form a single nucleus. Then the egg cell can develop into a larva in the usual manner.

From the very beginning, the female egg contains the male sperm, keeping it in reserve until it is needed. If the egg is not fertilized by a male, the "substitute male" stored in the egg goes into action to ensure the production of male offspring.

Sometimes women (never men) ask me whether human beings could reproduce by parthenogenesis. The question has ominous implications. If humans could reproduce by parthenogenesis, then henceforth humanity could dispense with men.

For years it was believed that parthenogenesis occurred only in the insect kingdom and among the lowest forms of animal life. Now we know that this is not true. The Amazon molly, for example, is a vertebrate. Moreover, zoologists have discovered that parthenogenesis occurs among three species of rock lizards in the Caucasus region and in Anatolia, as well as among whiptail lizards, also known as racerunners, in the region between China and the Caspian Sea. These species consist solely of females. Unlike the Amazon molly, the female lizards do not mate with the males of other species to activate the development of the embryo. Instead, they simply lay unfertilized eggs. Only females hatch from the eggs—or nothing at all. Examining eggs that never hatched, zoologists discovered that they contained dead and crippled male embryos. Among these lizard species, males are aborted before they can be hatched.

Thus clearly parthenogenesis does not represent a primitive method of reproduction which pre-existed the "invention" of the male. Instead, parthenogenesis developed in species already possessing both males and females. For some unknown reason, the males proved harmful to their species or simply superfluous. (By the way, the scientists who are working to discover the reason for this are all men!)

In the 1960's, scientists succeeded in artificially inducing parthenogenesis in vertebrate species.[1] The experimental animals were turkeys, rabbits, and mice. The most successful experiments were conducted by Soviet researchers, who in 1970 attached metal electrodes to the ovaries of mice and stimulated the eggs with electrical impulses. The mice were anesthetized and their abdomens surgically opened to permit the attachment of the electrodes. Shortly after the electrical stimulation of the eggs, segmentation of the germ cells occurred. However, after one week the embryos ceased to develop—presumably as a result of the surgery and the constant observation. At present, there appears to be little chance of producing living mice by this method. Thus there can as yet be no question of inducing parthenogenesis in human beings.

However, as soon as scientists begin to experiment in a new field, they open up many avenues of research. Male scientists must secretly have been alarmed at the thought that one day in the far-distant future, parthenogenesis might be artificially induced in human females, and the human male might become superfluous. Thus they have begun to experiment with "male parthenogenesis," or androgenesis. Sucking the nucleus from the female egg cell, they replace it with a male sperm cell, thus removing all female chromosomes from the egg and replacing them with male chromosomes. Turning the tables on women, male scientists are attempting to create a world in which males dominate females to an even greater degree than they do now. What a pity for them that even in their golden age of male dominance, women would still be needed to produce eggs and thus could not be done away with altogether!

Moreover, even an organism that grows from an egg containing only male chromosomes will inherit some traits from its mother. Besides the chromosomes, the cell plasma of the egg

Two mute swans declare their affection by curving their necks and placing their foreheads close together. Viewed from the side, the two swans form a heart. Young swans perform this gesture to signal that they are engaged. They become engaged in December and January, many weeks before they are sexually mature and capable of mating. Mute swans are monogamous and remain faithful to each other all their lives. A mute swan couple breed in isolation, not in a large flock like the greylag goose. Thus mute swans have no opportunity to be unfaithful or to exhibit nymphomania, as do many greylag geese.

helps to determine the traits of the offspring. This is why the mating of a stallion and a she ass produces a hinny, whereas the union of a male ass and a mare produces a mule. In both cases, the offspring resemble the mother more than the father.

People often forget this basic law of heredity and mistakenly believe that children inherit half their qualities from their mother and half from their father. In reality, the law that applies to horses and donkeys applies to human beings as well. All children inherit the bulk of their hereditary traits from their mothers.

Other scientific experiments with egg and sperm cells help to illustrate this fact. For example, scientists have removed from the egg cell the nucleus containing the female chromosomes, replacing it with sperm cells of other species. That is, they have attempted to induce androgenesis between animals of different species. This process is called merogony.

Alert readers will notice that this kind of experiment can be used as a weapon in the laboratory war of the sexes. Experiments in merogony could serve as a weapon against women. If it were possible to produce human beings by transplanting human sperm into the egg cells of animals, then women would be spared the trouble of carrying and giving birth to children. However, by the same token, women would have become unnecessary to the perpetuation of the species. It would indeed be a man's world.

The only significant experiments in merogony have been conducted on newts. The sperm of the palmate newt, the Alpine newt, and the crested newt were injected into the eggs of the smooth newt. However, the resultant embryos never developed to the point of being able to survive independently of the mother. The less closely the two newt species were related, the sooner the embryos ceased to develop. The genetic information contained in the male chromosomes did not harmonize with that contained in the egg plasma. As a result, the eggs produced deformed creatures incapable of life.

Thus, so far, all attempts to rob women of their biological *raison d'être* by means of genetic manipulation have failed. The geneticists' battle of the sexes is at a standstill.

At this point we must ask the question: What purpose do males serve? Apart from the various Amazon species and the species in which the ratio of female to male is one thousand to one, the "lords of the creation" comprise half the world's animal population. They consume half of the available food supply. What do they offer in return?

Do they help raise the young? In the animal kingdom, there are few species in which the males help to rear the young. Do they offer protection against enemies of other species? Such protection as they do offer is usually quite ineffectual. Do they offer protection against enemies of the same species? If there were no males, then no males would be needed to offer protection against them. Thus the only possible justification for the existence of males is the advantage offered by sexual reproduction. In what does this advantage consist?

The Amazon molly shows how monotonous the world would be if all offspring were exact replicas of their mothers. But does the uniqueness of each individual organism really constitute a biological advantage? Isn't the idolatry of individuality merely a private prejudice of human beings? And would it not in fact be more advantageous if each species consisted of the single most efficient "model," which could be reproduced a million-fold? Human beings admire and enjoy individuality. But individuality is merely a by-product of evolution, not its driving force.

We have already noted that the sexual exchange of genetic information enables certain bacteria to become immune to human drugs. Thus sexuality accelerates the evolutionary development of a species. However, taking into consideration mutation rates, size of population, and the effects of natural selection, biomathematicians have calculated that an animal population must reach one hundred million before sexual reproduction begins to accelerate the tempo of evolution. When the population is less than one hundred million, the existence of males slows rather than accelerates evolution.

However, sexual reproduction does offer one advantage over other methods of reproduction. If a beneficial mutation occurs in a species that reproduces by parthenogenesis, it takes a long time for the new trait to spread throughout the species. Very gradually the "old editions" die out and are replaced by the "new, improved edition." Not until the entire species consists of "improved editions" will it be in a position to benefit from another advantageous mutation.

On the other hand, given a population of sufficient size, sexual reproduction enables a species to acquire two or more advantageous traits at the same time, without waiting for the bulk of the species to become extinct. So stated, the advantage of sexual reproduction may appear minimal. However, in reality the advantage is so great that, painful as the fact may be to some devotees of the women's movement, very few species have failed to develop males.

# CHAPTER 2

# Male and Female in One Person

*The Creation of the Male*

Alligator Reef, a small coral island in the Florida Keys, is a little paradise complete with palm trees, sand, waves, and azure sky.

The moment the waters of the Caribbean close over his head, a scuba diver finds himself in another paradise, a pleasure-garden out of the Arabian Nights. At a depth of between ten and twenty-six feet, the diver will encounter broad fields of brilliantly colored sea anemones swaying to the rhythm of the waves. The water seethes with fiery red corals and sponges, and emerald green plants entwine colonies of sea urchins. In deeper water, anemones shaped like shaving brushes unfurl their crowns until they resemble Oriental baldachins. Five particolored fish of the species *Serranus subligarius* dance in a fairy ring at the edge of a reef. The largest is a fierce-looking creature six inches long, its suit of scales a gleaming orange dotted with dark blue spots and streaked with white stripes that match the white tips of its fins. The fish is hovering near another fish whose coloring is far more subdued. No, there is no doubt: The larger fish is a male, and he is courting a female. We can call him Paul.

Lining up parallel to his Pauline, Paul begins to vibrate like a drum. Simultaneously the female expels her eggs, which float down through the water looking like tiny soap bubbles. Paul looses a shower of milky fluid, fertilizing the eggs.

Next we see an uncanny apparition. Seconds after the eggs are fertilized, the brilliant orange fire of Paul's body seems to flicker and go out. The dark blue spots grow larger until the fish's entire body has turned indigo speckled with purple. Black borders dim his gleaming white fins. Meanwhile, in a flash Pauline has decked herself out in the same fiery raiment her lover wore just a moment before.

Pauline now looks and behaves like a male, circling around Paul and performing all the rituals of courtship. Suddenly Paul expels the eggs and Pauline sprays them with sperm.

Thus the transformation of the two fish was more than skin deep. Immediately after the eggs were fertilized, the male changed into a female and the female into a male. *Serranus subligarius* is male and female in one, and it can change roles in the space of a few seconds. It is a hermaphrodite like the legendary offspring of Hermes and Aphrodite, who was both man and woman and became the ardent lover of both the nymphs and fauns.

In modern times, underwater spear-fishermen are decimating the fish population of the coral reefs. The ability of *Serranus subligarius* to change its sex helps to preserve the species. When two fish meet, they can mate even if both are of the same sex, for either one can instantaneously change to the opposite sex. In fact, by exchanging roles, the pair can mate twice. Moreover, if a *Serranus subligarius* fails to meet another of its kind, it can spawn and then immediately change sex and fertilize its own eggs. Thus the female can act as her own male.

This fish reveals the following fact about the history of evolution: the existence of males may prove beneficial to a given species, yet it is by no means necessary that the two sexes be embodied in separate individuals. In fact, a few species seem to find it advantageous to live their lives as hermaphrodites.

*Serranus subligarius* is not the only fish capable of changing sex. There are many hermaphrodites in the families of the sea bass, wrasses, and sea breams. However, these fish do not change sex nearly as rapidly as *Serranus subligarius*.

The sea bass family includes the giant grouper, which can attain a length of thirteen feet and weigh half a ton. At around the age of three, when they reach sexual maturity, most sea bass are females. Depending on the species, the females change into males somewhere between the ages of five and ten years. The sexual transformation of an animal from female to male is known as protogyny.

Among many sea bass, the male is simply a female which has grown older. These species consist only of young females and older males; there are no young males or older females.

The female sea bass passes through various transitional phases before becoming a male. Of the more than one hundred and fifty species of sea bass, some species pass through phases when they are neither male nor female. Others are "simultaneous hermaphrodites." That is, for brief periods they are male and female at the same time. The length of time a fish spends in these transitional phases varies from species to species. In *Serranus subligarius* the transition takes only a few seconds, and the fish retains its ability to change sex for many years.

The wrasse family of fishes is even more bizarre than the sea bass. In certain species, there are two types of males. The first type is a male which was a female when it reached sexual maturity and which spent several years as a female before changing into a male. The second type is a "primary male," which was a male from the very beginning. Thus at one point in the history of the creation of the male, the "real male" developed alongside the hermaphrodite.

The two types of male wrasses hold different positions in the social order. The primary males are "he-men": large, strong, boldly colored, aggressive, and skilled at building nests on the ocean floor. The males that derive from females are smaller and retain some of the female's drab protective coloring. Three dark horizontal bands on their sides serve as signals of their subordinate rank. The signals pacify the aggression of the primary males. Occasionally these effeminate males try to build nests but they do a poor job. Theirs is the humble role of acting as second-in-command to the primary males.

When the master of the house is at home, his second-in-command hovers submissively near the nest. It is his task to protect the nest from robbers while the master is away visiting one of his three, four, or five other nests. To be sure, while the primary male is away, a female may swim up to the nest. In this case, the "servant" shows that he is as much a male as his master.

Soon after their development, primary males assumed a dominant role. They ruled by virtue of their aggression, their physical strength, and their domineering behavior. Only by thus arrogating power could males raise themselves from their lowly position as mere mating partners. Whenever males have failed to achieve a higher position than that of fertilizing female eggs, they have led a truly wretched existence. Immediately after mating, the males of many spider species serve as dinner to their mates. And once the drones have gone on their wedding flight with the female bees, they can do nothing but waste good food; thus they are driven from the hive to die in exile, or are simply stung to death.

We have noted that among the sea bass, the males develop from the females. Does this fact imply that the male represents a higher or more advanced life form than the female?

By no means does it imply this. In other species, the female develops after the male. For example, in the tropics lives a snail known as *Achatina achatina*, which looks like a giant version of the common edible snail. This snail sometimes weighs as much as a pound and carries around a house eight inches tall. The edible snail is a simultaneous hermaphrodite; the two partners fertilize each other simultaneously. *Achatina achatina*, on the other hand, becomes a male somewhere between the age of six weeks and one year. Later it develops into a female. Thus in this case the female represents a "more advanced" form of the male.

Sexual transformation from male to female is known as proterandry.

In other species, sex depends on the size of the body rather

than the animal's age. For example, everyone who has vaca-
tioned on the North Sea has encountered the species of nereid,
or marine worm, known as *Nereis virens*. As long as there are
fewer than twenty segments in its body, this creature remains
a male and produces sperm. When it grows longer than twenty
segments, its body begins to produce eggs instead of sperm.
Thus in this species, the young are all males and the larger
adults all females. If we cut off part of the female's body so
that it contains fewer than twenty segments, it promptly becomes male again. Moreover, if two females are placed in a
preserve jar half filled with sand and left together for two or
three days, the smaller of the two females will change back into
a male so that it can fertilize the female's eggs. Love casts a
spell on it so that it becomes capable of magically changing
its sex.

Another creature besides *Nereis virens* appears to make use
of Circe's magic wand. In most species, the accidental configuration of chromosomes at the moment of fertilization determines
the sex of the offspring. But in one species, the larva is neuter.
It has the capacity to develop into either a male or a female.
Only if it is touched by a female will this neuter creature turn
into a male.

Scuba divers have observed this bizarre animal on the rocky
floor of the Mediterranean. The female is green and about the
size of a dill pickle. From one end projects a proboscis a yard
long that opens into two terminal processes resembling leaves
on a stalk of rhubarb. This female is an echiurid called a green
bonellia.

Unlike the mother, the larva is tiny. At first it just drifts
around with the current. If it accidentally touches the proboscis
of its mother, it clings to the proboscis; then several days later
it changes into a male only a few millimeters in length. At this
point the male stops growing and spends his days as a sort of
"pimple" on the body of his gigantic mate, gradually making
his way into her intestines and finally into the oviduct, where
he is destined to live. Here he becomes a part of his mate, living
parasitically on her "fortune," with nothing to do but fertilize
the eggs as they pass by like parts on an assembly line.

As many as eighty-five of these dwarf males have been found inside the body of a single female bonellia. After her eggs develop into larvae, the mother touches all her children with her magic wand, transforming them into husbands with whom she will live in perpetual polyandry.

However, some of the larvae may drift around in the sea for a year without encountering a female of their species. If this happens, they grow into the very thing that they were seeking. Larvae which find no female to transform them into males become females themselves.

The female green bonellia looks and behaves like a magical plant. This creature resembles a pickle and possesses a proboscis ending in leaflike processes. Her body is over a yard long. When sexless young bonellias touch the female, they are changed into tiny males. In the picture, three males resembling buttons are making their way along the proboscis to enter the female's intestines.

No science-fiction writer musing about distant planets could have conceived a more bizarre sexual relationship than that between the green bonellia and her mate right here on earth. However, bizarre as they may appear, the mating habits of the green bonellia are highly efficient; for they enable the bonellia to regulate population growth and to prevent the development of a surplus of males and a shortage of females.

The tiny male bonellia, the parasite living in the ovaries of his wife, represents the very quintessence of maleness. His body is designed solely to facilitate the production of sperm. Nothing about it resembles the female's physique and functions.

It sounds incredible: During the first stage of its life, the green bonellia is so constituted that it can become a male or a female with equal ease. Yet later the two sexes become so distinct in size, shape, and bodily function that they hardly appear to be members of the same species. The bonellia exemplifies extreme sexual dimorphism. That is, the male and female have completely distinct forms, and each is adapted to fulfill its specific sexual functions.

It might seem logical to assume that in most species, male and female would exhibit completely different structure and behavior. However, among the higher animals it is apparently impossible for the sexes to be very distinct. The more complex animals pass on a wide variety of hereditary traits. In this case, relatively few genes can be devoted to sexual traits. Thus among the higher species, male and female are similar in many respects.

The greater the similarity between male and female in a given species, the more feminine the male will be and the more masculine the female. Moreover, the greater the similarity, the more difficult it is for males and females to tell each other apart. In a later chapter I will discuss species whose members do not at first know whether they are male or female.

In the animal kingdom, the distinction between hermaphrodites and animals with only one sex is not clear-cut. For example, a frog appears to have only one sex; yet we cannot say that it is totally male or female. If we castrate a male frog, he does not turn into a eunuch as a man would, or into an ox as a bull would. Instead he becomes a female. Moreover, a female frog can simply change into a male and stop producing off-spring, if food is scarce, and her young, if born, would have nothing to eat.

Human males and females are not as different as is often assumed. Many cases are known of people who have been born men and gradually evolved into women, and vice versa. To be sure, as a rule such a person must undergo surgical alteration of those few parts of his body which retain the traits of his original sex.

Human hermaphrodites manufacture almost equal quantities of male and female hormones. However, at any time they may begin to produce more of one or the other.

Pseudo-hermaphroditism can occur among humans. In such cases, the person possesses the reproductive glands of one sex and many physical traits of the other. In the 1960's, we all read headlines about highly accomplished athletes who looked like women but were actually men.

Homosexuality represents, in a milder form, the lack of a distinct sexual identity. For brief or prolonged periods a man may experience the emotional responses of a woman and thus feel sexually attracted to men. A lesbian woman feels the responses of a man. Thus homosexuality might be characterized as a sort

We still know little about the mysterious light emitted by some insects. Male and female fireflies signal each other with a series of blinking lights in patterns characteristic of each species. Using lights, they can arrange a rendezvous. But why do the larvae of some insects glow? This picture was photographed by the light of the larvae themselves. It shows the larvae of the South American click-beetle. Each of the ten segments of the larva's body acts as a lamp that emits a lemon-yellow light. Larvae are sexually inactive. Are the organs that emit light in the adult beetle so complex that they must be fully developed in the larval stage? No, for the adult beetle has only three lights, and they are not located in the same places as the lights of the larva. Do the larvae use the lights to frighten away enemies? If a human being picks one up, it glows more brightly. This fact seems to indicate that the light may be intended to frighten enemies. However, frogs, which are the beetles' principal enemies, find their prey by the light they emit. Sometimes a frog will eat so many of the beetles that its abdomen begins to glow.

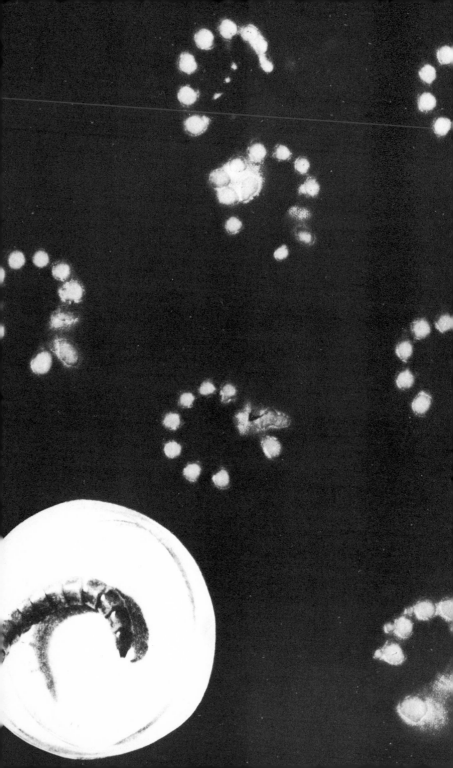

of partial hermaphroditism. The homosexual experiences the emotions of one sex while he is trapped inside the body of the other.

When a young child whose hormones have a tendency to be unstable is exposed to certain environmental influences, he may easily cross the emotional borderline between the sexes and begin to respond to the world as if he were a person of the opposite sex. In this case, he may become a homosexual. I will discuss this subject further in a future book concerning the relations between parents and children.

For now, I will simply point out that many animals must also deal with the problem of homosexuality. This fact proves that homosexuality does not represent a form of immorality; nor is it an abnormal or disease phenomenon. On the contrary, homosexuality is a common phenomenon in nature.

# PART TWO

## Signals That Attract
## and Repel

# CHAPTER 3

## Love at First Sight

*The Discovery of the Bonding Instinct*

Kai was a sturdy, pleasant young man of nineteen who was now going with his sixth girl friend. Although the girls he had liked were different in many respects, they all had one thing in common: They had round, soft, gentle-looking faces. Whenever Kai saw a girl or a picture of a girl who fit this description, he felt attracted to her.

Kai himself did not in the least resemble the women he admired. On the contrary, he had a stern and energetic-looking face with a determined chin. His women friends all admired his austere features.

People who work for dating services and matrimonial agencies know that frequently men are attracted to women with facial structures opposite to their own, and vice versa. However, dating service agents were not the first to discover this fact. It was accidentally discovered by a scientist investigating the response of human beings to weather.

Several decades ago, bioclimatologist Manfred Curry[1] noticed that a certain type of person became irritable and fretful during warm weather, whereas a different type reacted negatively to cold weather. Furthermore, he discovered that people adversely affected by warm weather were often married to people adversely affected by cold.

Depending on whether they were sensitive to warm or cold weather, Curry classified people as belonging to group W or group C. Those belonging to group W tend to have round heads, full cheeks with dimples, large eyes, smooth, convex foreheads, and full lips. Usually the corners of their mouths turn up.

The C group are just the opposite. They have long, angular heads, taut cheeks, flat, furrowed, "intellectual" foreheads, and narrow lips which they tend to press tightly together. The corners of their mouths turn down.

Most people are hybrids and possess traits of both groups. For example, a person who basically belongs to the W group may have eyes characteristic of the C group. Nevertheless, a single rule of thumb applies to everyone, including "hybrids": The more pronounced the group W traits of a person, the more pronounced the opposite traits will be in his or her spouse. People whose faces exhibit W and C characteristics in equal measure tend to marry people of the same type. Readers of this book might make a sort of game out of analyzing couples they know to see if they obey this rule.

Many people have observed the fact that there are different facial types and that persons of opposite types tend to be attracted to each other. However, it has not yet been scientifically proved that this is true. Undoubtedly we all know people who seem to be exceptions to the rule.

Kai was an exception too. At the age of twenty-three, he married a woman quite different from the women he had always admired. His wife was a prosaic sort of person, and the two of them appeared to have a dull marriage.

Many computer dating experts have despaired over the tendency of men to marry women very different from those they had hitherto admired. I believe that people often marry someone who is not "their type" because they do not understand the significance of the attraction between persons of opposite physical types.

"Love at first sight," or the instinctive attraction to a person of the opposite physical type from oneself, springs from an unconscious feeling of liking or affinity. Suddenly two people experience a sense of kinship that defies rational explanation. The ancient Romans described the experience as being shot with Cupid's arrow.

An instinctual law decrees that opposites attract. As a rule, the attraction is mutual. One partner may at first make all the advances, while the other performs various preliminary rites, pretending to reject his advances and putting him to various tests. In the end, the courted partner will probably respond in kind.

This instinctive feeling of sympathy or affinity forms the emotional basis of the bonding instinct. What exactly is the bonding instinct?

The discovery of the bonding instinct in 1965 was an epoch-making event in the history of animal behaviorism. In my view, this discovery sheds more light on human behavior than all the treatises on aggression written during the past fifteen years. Unfortunately, most people still know nothing about the bonding instinct. And despite the fact that all social behavior ultimately derives from this powerful drive, most animal behaviorists have hitherto ignored its importance.

The following statement of Helga Fischer suggests the fundamental role the bonding instinct plays in animal behavior: "Today it is still widely accepted that all social behavior derives from the 'instincts' of sexuality, flight, and aggression. This view has proved false in the case of the greylag goose. The social relations of the greylag goose are rooted in another instinct, the bonding instinct. The work of Carpenter, Washburn and DeVore, Altmann, and Goodall suggests that probably the same instinct plays a role in the behavior of other animals which maintain social ties—for example, primates." [2]

It is the purpose of this book to depict the bonding instinct and the role it plays in the mating and marital behavior of animals and human beings.

As the quotation from Helga Fischer makes clear, the bonding instinct is unrelated to sexuality. Moreover, the passage suggests that social ties, including the bond between male and female, are not a consequence of the sexual drive but of another, independent instinct.

The discovery of the bonding instinct, which Fischer described in a few clear-cut phrases, may shed some light on the problems of contemporary marriage. Many marriages are breaking down and many children are being raised in homes where the parents feel no love for each other. To some degree, the crisis of modern marriage may have resulted from a misevaluation of the forces which really bind people together. Ignorance of what makes for an enduring relationship has led many people to look for the wrong kind of partner.

According to Freudian psychoanalytic theory, all human behavior is governed by the sexual instinct. Moreover, modern social reformers frequently view freedom from sexual repression and inhibition as the key to a better world.

Even the writers of a modern, twenty-volume encyclopedia seem confused as to what really constitutes the "tie that binds." Describing the phenomenon of sexuality, one volume states: "The sexual instinct, which is closely linked to the sexual organs, makes possible the reproduction of the species and also leads to love and the formation of enduring ties." [3] Small wonder that many people assume that sexual attraction represents the only possible basis for an enduring marriage.

In this book I will show that the sexual instinct and the bonding instinct, based on sympathy, are two distinct forces. At times the two may coexist and influence each other; but by no means does the sympathy bond derive from sexuality.

The bonding instinct acts as a socializing force. Pure sexuality does just the opposite. It brings male and female together to mate, but this relationship lasts for only a few minutes and is primarily aggressive in character. After mating, two animals are as indifferent to each other as they were before. When divorced from the bonding instinct, sexuality may have the character of prostitution or rape.

Sexual desire centers on areas of the body below the neck, sympathy on the face. The sexual drive causes two creatures to satisfy a physical need. The sympathy bond causes them to seek each other's companionship.

To return to our original discussion, what leads many dating services to arrange unsuitable matches, and why did Kai enter into an unhappy marriage?

Before his marriage, Kai unconsciously allowed his instinctive sense of sympathy, or natural affinity, to govern his relationships with women. But when he began to look for a marriage partner, he mistakenly assumed that sexual attraction guaranteed a happy and loving marriage, and sought a wife whose charms were mostly below the neck.

We have noted that opposites attract, so that people belonging to group W instinctively like members of group C. Persons of opposite physical types may also be sexually attracted to one another. However, the types which attract each other sexually are not necessarily identical to the types represented in the W and C groups. In other words, sympathy does not always go hand in hand with sexual attraction. A man belonging to group C may be charmed by the face of a woman in group W, and yet feel sexually indifferent, or even repelled by her. By the same token, a man or woman may be sexually attracted to someone whose face he or she finds unappealing. In a story by Heinrich Böll,[4] the male protagonist covers the faces of prostitutes with a towel before making love to them.

Frequently, persons contemplating marriage take a bewildering number of things into consideration. They may worry about money, about whether their marriage will heighten their social prestige, about which of the partners will "wear the pants" in the family, and whether both partners want children. However, compatibility in all these areas does not guarantee a happy marriage; nor does sexual attraction. If their marriage is to be successful, there must be a strong personal bond between the two partners.

When two people marry on the grounds of sexual attraction, they are both likely to be possessive, may easily come to dislike each other as people, and may both be prey to jealousy. At bottom, jealousy springs from a sexual inferiority complex coupled with the desire to make one's partner feel guilty.

It is frequently said that love never lasts and that when

love has died, the sense of duty should compel married couples to remain faithful. Unfortunately, when the love has gone out of a marriage, few people can resist the temptation to stray. Scores of private detectives make a living largely from tracking down evidence of infidelities. Nevertheless, it is inaccurate to say that love never lasts. True, sexual attraction cannot endure; but a strong sympathy bond can endure and keep a couple together.

The bonding instinct represents a stronger drive than the sexual drive. As the sexual relations of animals reveal, passion may for a time be stronger than the sympathy bond, but it is of shorter duration. If the sympathy bond is strong to begin with, occasional infidelities will not destroy it. Infidelity results in divorce only when the sympathy bond is weak. If it is strong enough, it will survive even when the two partners have lived together for years and know all each other's faults. It can even survive the death of one of the partners.

The picture shows two black-headed gulls during copulation. The mates never dare to look each other in the face. In autumn and winter, black-headed gulls closely resemble other gull species; but during breeding season they wear black masks that are as frightening to other gulls as demon masks are to primitive peoples. While the gulls are in the breeding colony, their masks keep all the other gulls at a distance. As a result, it is difficult for the birds to get close enough to each other to mate. First the male selects a nesting place in the grass at the edge of the dunes and then calls to attract a female. If an unattached female hears his call, she will land beside him. For a moment the two look at each other as if they were enemies. Then one of the birds turns so that the two of them are standing side by side. Now that they are facing in the same direction, their frightening masks no longer terrify the birds but are turned on empty space. Finally both birds assume an imposing posture that suggests threat behavior. However, to make it clear that they do not mean to attack each other, they turn their heads away from each other. Thus the courtship ceremonial of black-headed gulls reflects the birds' aggressiveness, desire to flee, and desire to remain together. These gulls are monogamous, but each year they are together only during the breeding season. While they are together, they never dare to look at each other's faces. If they do so, the result is quarrelling or divorce.

The sympathy bond, like other forms of instinctual behavior, developed early in evolutionary history. To find its roots, we must turn to the animal kingdom.

Most animals could not exist and reproduce without "love at first sight." Their lives are short and they have a limited learning capacity. Only instinct can tell a mayfly what its sexual partner should look like. It can never look in a mirror, see its own image, and look for another creature that resembles itself. And of course, it does not even know why it is supposed to mate.

Thus at birth, most animals possess a sort of "passport photo" of their sexual partners. Suddenly they see something that attracts them so strongly that without knowing why, they must at all costs run, fly, swim, or crawl over to it—just like a human being who falls in "love at first sight."

As a rule, the "passport photo" an animal is born with is not a complete portrait. To instinctively recognize all the details of a shape, an animal must possess a highly complex nervous system. Small creatures like insects cannot register such complex impressions. Thus to recognize a sexual partner, many animals respond to several striking symbols.

For example, a love-starved male housefly will fly over to anything that is (1) the size of a fly and (2) dark in color. Everyone has observed the behavior of flies in an animal pen. A male fly will not only alight on other flies, male as well as female, but also on the heads of screws or small heaps of animal dung. He will make many mistakes before he finds a female; but in a pen full of flies he is bound to be lucky sooner or later.

Mating is far riskier for the bedbug, for mating is sometimes tantamount to murder. The male's sexual organ resembles the sharp curved blade of a scimitar. He does not attempt to insert it into the female's genital opening but treacherously stabs her in the back and releases the sperm into the bloodstream, which carries it to the reproductive organs. Sometimes the male may actually kill the female; but as a rule her wounds heal. By counting the number of scars on the back of a female bedbug, we can tell how often she has mated. Even males have these scars on their backs, for bedbugs are unable to distinguish between males and females, and often a male will "mate" with another male. The male bedbug will stab anything that is (1) the size of a bedbug, (2) dark in color, and (3) flat in shape. Of course, the sperm injected into males produces no offspring.

Among higher animals, the "passport photo" of the sexual partner is designed to prevent such cases of mistaken identity as occur among bedbugs. The "photo" enables the animal to recognize members of the same species and also to distinguish between male and female.

For example, tropical and subtropical fireflies searching for mates in the darkness recognize their partners not by their physical appearance but by a symbolic system of blinking lights.

Various species of fireflies inhabit the same region of the southern United States. The tiny skywriters have developed an elaborate signal code to prevent cases of mistaken identity. If it were not for this code, male fireflies might mistakenly mate with females of other species.

Every 5.7 seconds, the male *Photinus pyralis* rises in the air and then drops again, outlining a series of little hills and valleys. Just as he reaches the valley, he switches on his yellow-green lantern and does not turn it off until he reaches the crest of the next hill. Thus he writes a large letter "J" in the darkness.

The male of a closely related species outlines much lower hills, and as he ascends the curve, emits three blinks of light. Another male belonging to the same genus, *Photinus*, flies in a straight line and blinks once every .3 second. A fourth male of this genus flies in a straight line and every 3.2 seconds executes an arc which he outlines in light. In a fifth species of *Photinus*, every 2.7 seconds the male's body lights up and cuts a zigzag pattern, writing six tiny letter "m's" against the night. A male firefly of the genus *Photuris* hovers in the air like a helicopter, begins to glow faintly, then grows brighter and brighter until, at the height of his performance, his lantern suddenly goes dark.

By no means do these six codes exhaust the repertoire of fireflies. There are some two thousand different species, and each has its own code of blinking lights. The codes vary in the color of the light, the shape of the light pattern, the time interval separating each "glow," the length of time the light persists, and the modulations of light intensity. In many respects, firefly codes resemble the beacon code used by ships at sea.

The female fireflies, or "glowworms," wait in the grass. When they recognize the code of a male of their own species, they respond by turning on a "landing beacon." If the male makes even a minor mistake while emitting his signal code, the female remains in the darkness and gives no answering signal.

Even minimal errors in signalling will prevent the male firefly from attracting a sexual partner. From observing mimes and actors, we all know that almost imperceptible changes in a person's expression can make us suddenly distrust someone we had hitherto liked and trusted. In the same way, an error in a male firefly's code alienates the female.

As a rule, the female firefly responds to the male's signal by emitting a brief blink of light. However, if the male is to recognize her, she must send her signal at exactly the right moment. The female *Photinus pyralis* must light her lamp exactly 2.1 seconds after the male's signal. In another firefly species, the interval is 2.2 seconds. Thus if the female *Photinus pyralis* lights up a tenth of a second too soon or too late, the male will fly right past her. As I have already stated, there are some two thousand different species of fireflies. The females of most of these species emit their signals between .2 and 4 seconds after the male has signalled. Thus the interval between .2 and 4 seconds is very "crowded." To avoid mistakes, the timing of the different species must be absolutely precise.

Female fireflies of one species of the genus *Photuris* are carnivorous. They exploit the fireflies' symbolic code of blinking lights to get their dinner. The females crouching in the dark grass wait for prey as well as for lovers. If a carnivorous female recognizes the signal of a male *Photinus pyralis,* she blinks her own light exactly 2.1 seconds later, waits for him to come to her, and then eats him. This female knows the signal codes of at least twelve firefly species and responds to the males of each species with the proper "password." She is the

The fireworks display of fireflies above a field in a subtropical region. The male of each species signals his arrival to the female, which is waiting in the grass. The signal consists of a pattern of blinking lights unique to each species. The text describes the various patterns in detail.

siren of the animal kingdom, who seduces males not with music but with light.

The behavior of this carnivorous female firefly reveals an important fact about the instinctual schemata which enable animals to recognize their sexual partners. Under normal circumstances, these schemata function quite efficiently. However, frequently animals cannot tell the difference between the real thing and a mimic. Thus a signal that normally helps them to function may lead to their destruction.

At times human beings fall into the same kind of trap and are deceived by the wiles of swindlers—for example, the kind of swindlers who promise marriage. When one of these treacherous gentlemen is brought to trial and confronted by the ten or twenty women he has swindled out of their money and their love, they are all likely to forgive him because "he's such a nice person." A mimic, a simulated signal, has led them astray.

Some people may feel that the analogy I have drawn between the behavior of fireflies and that of human beings is rather farfetched. However, I am not attempting to apply to human beings facts we have learned about animals, but rather investigating what underlies behavior patterns common to both. Identical behavior implies the operation of identical laws. The protein composing the bodies of men and animals is made up of the same basic elements, amino acids. In the same way, the instincts and behavior patterns of men and animals have certain elements in common. The sympathy bonding instinct is one such element. Moreover, we all learn more easily from observing others than from observing ourselves. The study of animals can contribute greatly to our understanding of human beings.

At night, when two insects cannot see each other's shapes, they must communicate with light. Thus the firefly reduces a complex form—its mate's body—to a simple symbol or abstraction. However, even in daylight members of many species recognize their mates by means of abstract symbols. For example, on Baffin Island in northeastern Canada, the herring gull lives in close proximity to three closely related species—the Iceland gull, the glaucous gull, and a gull called *Larus thayeri*.

The four species resemble each other so closely that human be-ings can hardy tell them apart. All four gulls are monogamous, but they mate only for the mating season and spend the rest of the year alone. Despite the close resemblance between the species, a herring gull never makes the mistake of mating with a gull of another species.

Clearly it is not the gray coloration of their wing-feathers that enables male and female herring gulls to recognize each other; for when scientists colored their wings a different color, the gulls still chose mates of their own species. How then does the herring gull recognize its sexual partner?

The four species of gulls on Baffin Island all have rings around their eyes. The rings are only one millimeter in width. Gulls of the same species recognize each other by the color of their eyes and eye rings. The herring gull's eyes are pale yellow; the rings around its eyes vary from medium yellow to light brown. The eyes of *Larus thayeri* are the same color as those of the herring gull, but the circles around its eyes are a paler yellow. The rings around the eyes of an Iceland gull are the same color as those of a herring gull, but its eyes are a darker yellow. When a herring gull encounters a bird of another species, it apparently feels as disinclined to enter into a sexual relationship as a human being would be to have such a rela-tionship with an orangutan.

A scientist can break down the social barriers between the different gull species simply by painting the eye rings more "attractive" colors. An alteration in the color of the eye rings can also destroy hitherto harmonious seagull marriages.

While experimenting with the eye rings of the herring gull, American zoologist Neal Griffith Smith [5] made some surprising discoveries. Before a female herring gull had found a mate, Griffith Smith painted her eye rings a darker color. The male herring gulls seemed to have no objection to the darker rings, and one of them became the female's mate. Then the zoologist darkened the eye ring of a male gull. The females refused to have anything to do with him, and he was unable to find a mate.

However, after mating the behavior of male and female

herring gulls is reversed. When the female's eye ring is painted a darker color, her mate rejects her. He cannot "forgive" her for the change in coloring, and the two mates quarrel and part. But if the male mate's eye ring is darkened, the female forgives him for his disfigurement and does not attempt to leave him.

It is difficult to explain these variations in the behavior of gulls before and after mating. Nevertheless, Griffith Smith's experiments reveal that signals designed to attract or repel mates produce different effects at different times. Depending on whether it comes from a male or a female, the same signal has a different meaning. Moreover, the same signal from the same partner produces different effects before and after mating.

These variations in the behavior of male and female gulls may lead us to wonder why men and women behave so differently before and after marriage. It is often said that once people are married, they relax and stop trying to impress each other. Frequently this relaxation produces disenchantment. For example, a man may feel disappointed when he finds out what his wife looks like without makeup. Makeup, hair styles, and modes of dress underline or alter the natural signals which attract or repel other human beings. As is the case with seagulls and fireflies, even slight variations in these signals may radically alter the effect. By changing her hair style or ceasing to pluck her eyebrows, a woman may appear to belong to a different "type" than she formerly did. As a result, she and her husband quarrel, although ostensibly they are quarrelling about something quite unrelated to the way she wears her hair.

The discovery of the bonding instinct gives us a key to understanding the psychological makeup of human beings and may even help us to avoid some kinds of marital problems.

In the silent film era, many American film stars were afraid to have their makeup changed or to appear in roles different from those which the public was accustomed to seeing them play. Many actors who attempted to play different "types" were promptly rejected by the public. In her last film, *The Woman with Two Faces*, Greta Garbo played two types of women. One was a "good," simple, energetic ski instructor. The other resembled the type Garbo most frequently played, the nightclub singer or vamp. In this film, the vamp was more villainess than heroine. Thus the audience could not identify with the "typical" Garbo role. As a result, the film was unsuccessful, and Greta Garbo gave up her film career.

Modern film stars also tend to play themselves or to play one role over and over, rather than taking the risk of playing characters of varying types.

Animals do not understand the meaning of signals designed to attract potential mates and repel members of other species. Instead, these signals act directly on an animal's senses. The signals can take many forms. Facial features, eye color (seagulls), blinking lights (fireflies), smells, tastes, sounds and melodies, gestures, and radiant displays of color in brilliant sunlight, can have an almost magical effect on potential sexual partners.

Let us look at a simple example from the insect world. Male fruit flies win the female's consent to mate by serenading her with a love song. Only if the male hits exactly the right notes will his plea be heard.

There are more than two thousand species of fruit flies, which swarm around rotting fruit. The tiny fruit flies are only some two millimeters in length. Many species resemble each other so closely that even an expert can tell the difference only by examining their internal organs under a microscope. However, a female fruit fly can recognize a male of her own species by the courtship dance he performs. The male dances not in front of, but behind the female. He follows in her footsteps, placing his feet a few thousandths of an inch from hers. At the same time he spreads one of his wings (not both) and makes it vibrate.

Up to this point, the courtship ceremonial of the various fruit fly species is identical. However, once the male begins to vibrate his wing, the female will soon know whether or not he is of her species. He belongs to her species only if he "sings" the correct song with his wing while he dances. If he is an acceptable partner, the female allows him to touch her with his proboscis and to mate with her. If he serenades her with the wrong song, she flies away or beats her wings to drive him away. Rape does not occur among fruit flies.

The love song of the fruit fly constitutes a sort of Morse

code consisting of humming notes. For example, the male fruit fly of the species *Drosophila melanogaster* hums a note corresponding to E on our scale. He emits this signal twenty-nine times per second. A closely related species, *Drosophila simulans*, hums the same note, but only twenty times per second. The male of a third species, *Drosophila persimilis*, also signals twenty times per second, but he hums a higher note corresponding to C on our scale. A fourth species, *Drosophila pseudoobscura*, hums the same note, but only five times per second.

Thus every species of fruit fly has its own secret code. Zoologists have discovered only two species with precisely identical "songs." However, one species lives in Europe and the other in North America. Thus there is no chance that a female may mate with a male of the other species and produce hybrid offspring.

However, scientists have succeeded in getting fruit flies of different species to mate. Gluing the wings of the male to his side so that he could not produce his humming song, they placed him near a female of another species and played a tape recording of the courtship song of the female's species. The female fell into the trap, mated with the male, and produced hybrid young.

The "vocabulary" of female fruit flies consists of only one signal, a rather long, loud humming sound. The meaning of this sound is "international," and it is understood by the males of all two thousand species of fruit flies. It means "No!"

Thus auditory as well as visual signals can trigger positive or negative responses in prospective mates. Even slight deviations from the "correct" form of a signal can change a favorable response to one of aversion or indifference. However, imitations of a signal can deceive the potential mate.

The way to a penguin's heart is through its ear.

Every year, more than one hundred thousand Adelie penguins from all over the southern hemisphere land on the coast of Antarctica to breed. The eyes of these penguins are designed for seeing underwater; on land they are very nearsighted. Thus to the throng of penguins all desperately looking for mates, one penguin looks much like another. No matter how handsome a male penguin looks in his "frock coat," his appearance makes no impression on members of the opposite sex. Instead, to impress the females the males raise their croaking, screeching voices in a great chorus resembling the braying of a donkey.

If a male penguin sees another penguin that looks as if it has not found a mate, he courts it by rolling a stone along the ground with his bill and courteously laying it at the feet of the other penguin. The courted bird can react in three ways to the gift of the stone. It can lean forward and screech in rage at the suitor. In effect this means: "You idiot, can't you see that I'm a male too?" If it beats the male with its stumpy wings, this means that it is a female which did not like the male's song. Or the female can accept the male's invitation to sing with him. If she chooses the latter course, she will begin to dance around him, meanwhile joining him in a lyrical and tender, if croaking, duet.

Human beings may wonder how the croaking of the Adelie penguin could produce an aphrodisiac effect. However, whether they are communicated by sight, sound, scent, or taste, signals which attract sexual partners are effective only when received by members of the same species. There are no universal standards of beauty, and no animal can understand what constitutes charm in another species.

Like penguins, human beings may have powerful emotional reactions to the voices of members of their own species. Human beings fall in "love at first hearing" as well as in love at first sight. Perhaps there is even such a thing as "love at first smell." In any case, we sometimes say, "He doesn't smell right to me," meaning that we instinctively distrust someone.

An entire industry keeps busy manufacturing products such as deodorants, deodorant soaps, and mouthwash, which are designed to abolish all natural odors and create an aromatic, absolutely sterile human being. Another industry, the perfume industry, creates imitation scents to replace scents lost in the deodorizing process and to stimulate favorable responses in other human beings. Regrettably, these imitation scents do not always provoke the desired response.

It is difficult to understand why millions of human beings allow advertising agencies to manipulate them into undertaking a fanatical crusade against the natural odors of their own bodies. In part, they may do so because they feel socially insecure or guilty. Insecure people need something on which to focus their insecurity, and guilty people need a scapegoat onto which to project their guilt. Odors can be blamed for everything; and by exorcising odors, people feel that they can exorcise their guilt or inadequacy.

Human relationships may to some degree be based on scents which trigger responses analogous to those produced by sights and sounds. Thus by interfering with natural odors, we may be affecting relationships. However, scientists have not yet fully explored the role which the sense of smell plays in our lives. Thus for now, we must turn to the world of animals if we wish to study the magical powers which fragrance can exert over living beings.

We all know what interest animals show in each other's scents. When two dogs meet, they always smell each other. Entering a veterinarian's waiting room, a male dog notices a female dog in the room. Trotting joyfully over to her wagging his tail, he first struts around her with the proud gait of a Lipizzan stallion, then begins to sniff at her rear end. Seconds later, his tail drops to half-mast. To save face, he sniffs here and there

Who is the fairest in the land? This question is very important to a male great white egret that is looking for a mate, for a female will mate with him only if she considers him extremely handsome. The taller the male is, the better chance he has of winning the female's admiration. Tall males possess large feathers, and females consider large feathers the most beautiful. (Witness the plumes human females have always liked to wear!) The picture shows a male egret during his courtship dance on the nest. Smaller males can improve their chances if they occasionally interrupt their dance to perform a graceful courtship flight. The smaller a male is, the more often he must fly about performing exhausting acrobatics. Thus the small males can make up for their physical shortcomings with beautiful gestures, which in the eyes of a female egret constitute charm.

at completely uninteresting objects and then trots away. She simply "did not smell right to him."

Because dogs pay more attention to smell than to appearance, a huge great Dane may make friends with a Scotch terrier.

When in heat, female whales leave a trail of scent behind them, the way airplanes leave a trail of vapor in the sky. Thus the male can follow a female he likes through the vast depths of the ocean.

On the African plains, a mother antelope smells her young as soon as they are born. Henceforth their scent enables her to recognize them. She will nurse only her own offspring and will refuse to have anything to do with other young antelope, which have a different smell. If one of her young dies, the mother will hold a "wake," waiting by the body until it begins to decay and she can no longer recognize its scent. This behavior of the mother antelope is not instinctive but learned. That is, her bond with her young develops only when she learns their scent.

The *Attacus atlas* moth is native to southeast Asia. With a wingspread of ten inches, it is one of the largest moths in the world, and its bright coloring and graceful shape make it one of the most beautiful. However, no one knows why the *Attacus atlas* should have evolved into such a large and beautiful moth, for neither the males nor the females show the slightest interest in size or beauty. The female sits on a branch and emits a fragrance designed to attract males. The males are quite indifferent as to whether the source of the fragrance is large or small, beautiful or ugly, round or flat, animate or inanimate. Moreover, the female is willing to mate with any male which responds to her scent.

Zoologists soaked a crumpled paper tissue with the scent of a female *Attacus atlas* and fastened the tissue to a branch.[5] Several males fluttered over and attempted to mate with the tissue. The real female, looking seductively beautiful, was sitting in a glass case right next to the tissue, but the males paid no attention to her. They were deceived by a mimic exhibiting one sympathetic signal possessed by the living female—her magical scent.

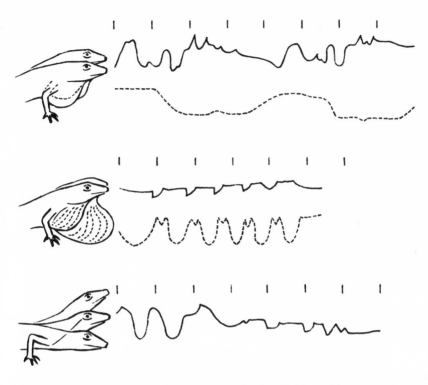

Lizards of the genus *Anolis* communicate by head-bobbing. The upper line shows the level to which the lizard raises and lowers its head. The dotted line shows the movement of the air sac at its throat. The divisions of the scale represent seconds. The upper picture shows the head-bobbing sequence by which the American chameleon communicates its species to other lizards. The second picture shows the sequence for *Norops auratus*. In the bottom picture, a male *Norops* nods lovingly to one of the females in his harem.

In other species, it is not beauty of form or voice or scent that arouses love, but rather beauty of behavior—in short, charm.

Various lizard species of the genus *Anolis* inhabit Central and South America. These creatures exemplify (at least to other lizards) the meaning of charm.

A *Norops auratus* lizard is sunning himself on a rock on one of the islands of the Pearl Archipelago in the Gulf of Panama. We can call him Little Fafnir, after the giant dragon of mythology. Then another *Anolis* lizard approaches the rock, twisting along the ground with the agility of a weasel. The stranger looks almost identical to Little Fafnir. One hundred and sixty-five known species of lizards belong to the genus *Anolis*, and only a zoologist can tell many of them apart.

Little Fafnir must learn the identity of the intruder. If the strange lizard belongs to another species, Little Fafnir will simply ignore him. However, if the newcomer is another *Norops auratus*, he constitutes a rival, and Little Fafnir must drive him away from his harem and his territory.

When the intruder is about five feet away, Little Fafnir leaps a foot or so toward him and then leaps another foot. The two miniature dragons face each other, only a yard apart.

Now the intruder begins to bob his head: Head up—head down—up, but not as high as the first time—down—halfway up—down—all the way up—halfway down—all the way up—halfway down—slowly all the way down. The strange lizard goes on signalling for 4.5 seconds. Then he repeats the same sequence. The second time he raises his head all the way, he inflates the air sac at his throat. His signals mean, "I am an American chameleon."

Little Fafnir, the *Norops auratus*, offers the stranger his own "visiting card." Every three-quarters of a second he lowers his head slightly, then raises it again. Between bobs he pauses to inflate his throat sac and immediately deflates it.

When he threatens, a male *Norops auratus* lizard can make himself look three times larger than his normal size.

Now both lizards know that they are of different species and thus are not rivals. The excitement is over. Henceforth Little Fafnir ignores the stranger, for members of other species merit neither hostility nor affection. The females in Little Fafnir's harem also ignore the intruder.

If, on the other hand, the head-bobbing code reveals that both lizards are of the same species, the recognition signal becomes a threat. Lining up parallel to each other, the two opponents inflate their throat sacs, raise the serrated combs on their backs, and inflate their bodies to make themselves look larger. By inflating their bodies, they can make themselves appear three times their normal size. At the same time they open their mouths and stick out their large fleshy tongues. In this impressive pose, both of them bob their heads up and down twice a second, simultaneously performing "push-ups," rising up on all four legs. This is the way lizards threaten.

Two equally matched lizards may spend an hour in these mock combats, threatening and trying to impress each other. As a rule, the lizard which feels inferior changes his color from green to a modest brown and leaves the field to the bright green victor. Only rarely do the two lizards engage in physical combat, lashing each other with their tails and leaping at each other fiercely.

The weaker sex is less concerned with avoiding bloodshed. Female *Anolis* lizards are somewhat smaller than the males. However, when they quarrel, they spend little time threatening each other. Instead they bite each other until the blood flows. The females are not as strong or daring as the males, but once battle is drawn, they are less inhibited about biting each other. In this respect, female lizards resemble the females of many other species.

Male and female *Anolis* lizards also bob their heads to identify themselves to each other. Head-bobbing can be used to threaten, to indicate species, or to communicate friendly feelings. In many animal courtship ceremonies, gestures connoting affection closely resemble gestures which express hostility.

Male and female *Norops auratus* lizards assume a threatening posture and bob their heads at each other like two sworn enemies. However, they suppress two signals of aggression: They do not open their mouths or stick out their tongues. In the same way, a human male will show off his strength to a woman much as he would show off to a male rival; but at the same time he makes it clear that he is trying to attract her, not to intimidate her.

When dogs are playfully fighting together, they must continually wag their tails and move their ears to indicate that they are only playing and do not intend to hurt each other. If a dog ceases to send the proper signals to his playmate, the mock battle will quickly turn into a real one.

There is a thin line between charm and aggression. More precisely, charm evolved out of aggressive, threatening behavior. Threat behavior, or signals of aggression, gradually changed into sympathetic signals that attracted a sexual partner.

Like animals, human beings communicate friendly feelings by small, involuntary gestures. I am not referring to smiles or to gestures like spreading one's arms wide to embrace someone. Both of these are conscious gestures which can be simulated at will. However, some reactions are involuntary. By analyzing photographs, Irenäus Eibl-Eibesfeldt[7] discovered that when a person meets someone he likes and trusts, both his eyebrows shoot up for about one-sixth of a second. This response is instantaneous and completely unconscious.

When the other person, also unconsciously, observes someone twitch his eyebrows, he is likely to think, "My, what a nice person!" Thus the meeting will get off to a good start. Smiles may easily be interpreted as insincere, ironical, or even malicious. However, the involuntary twitching of the eyebrows conveys an unequivocal message of friendship.

We often hear it said that "First impressions are best." In reality, first impressions are often deceptive; yet people frequently judge one another in terms of vague feelings they experience at their first few meetings.

Human beings invent justifications for everything they do and think. Reason does not function very rationally. We use our minds to justify what our feelings have already led us to believe. Our minds are not independent, but are the tools of our emotions. There is only one way that we can bring our feelings and instincts under rational control: We must learn to understand the power which instinct exerts over reason. Only when we understand our instinctual impulses can we bring them into harmony with the world of fact.

I once knew a highly intelligent seventeen-year-old boy who was having a difficult time in school. Little folds at the corners of his mouth gave him a slightly mocking expression, and his large, dark eyes, almost hidden beneath long locks of hair, looked rather sleepy. Most of his teachers considered him "impudent, stupid, and uninterested in schoolwork," and wanted to expel him from school. They judged everything he did in terms of his physical appearance. Thus even the fine papers he wrote could not alter their poor opinion of him. In the same way, people often perceive camels as being "stuck-up" because they hold their noses high in the air.

I talked to the boy's teachers, trying to explain why they instinctively disliked him. Once they understood their behavior, they were able to establish a good rapport with him.

Our emotions enable us to understand some things better than we could by the use of reason alone. Nevertheless, our feelings are frequently deceptive. Before judging anyone, we must ask ourselves what kinds of signals we are responding to and whether these signals are an accurate reflection of the person's intentions.

When we encounter another human being, our senses are bombarded with impressions. We receive visual and auditory signals, smell scents, and observe gestures. No doubt we receive many signals of which we are completely unaware. Some animals are sensitive to only one signal, whereas others can receive two signals successively or simultaneously. Human beings, on the other hand, are sensitive to a multitude of signals. Only when we understand all the signals which influence the unconscious mind can we truly be masters of ourselves. By learning to understand our instinctive responses, we can untie the Gordian knot—the tangle of instinct and reason—and use the string, like Ariadne's thread, to guide our behavior in the labyrinth of our emotions.

# CHAPTER 4

# Misguided Behavior

*Mimics Trigger Inappropriate Responses*

Some years ago, an electrical transformer station was erected near the fever swamps of the Brazilian town of Santos. When technicians turned on the power, millions of the mosquitoes which carry yellow fever came flying toward the station and were roasted to death on the hot machinery. A bulldozer had to keep clearing away the heaps of dead mosquitoes. Yet masses of mosquitoes continued to darken the sky, all flying toward certain death.

When in flight, the females of this mosquito species emit a humming sound consisting of between five hundred and five hundred and fifty vibrations a second. During mating season, the males fly toward the humming sound. They respond to anything which makes this sound as if it were a female of their species. Their response is instinctive; it is programmed into them before birth.

The transformers at the new electrical station made exactly

the same sound as the female *Stegomyia aegypti*. The male mosquitoes took this monstrous technological creation for a huge superfemale and were lured to their deaths as once sailors were lured by the song of the sirens.

Human technology has so altered the world that the natural instincts of these mosquitoes led them to disaster. Nor are mosquitoes the only animals to have suffered from technological advances.

The Eastern wapiti, a larger and more majestic cousin of the European red deer, is native to North America. During breeding season, the stags do not bell like European stags, but whistle.

One day a new type of electrical locomotive was installed on the railroad west of Lake Winnipeg. Quite by chance, the new locomotive made the same whistling sounds as the male wapiti. The deep tooting sound of the steam locomotives had always driven the deer away from the trains; but the stags attacked the new locomotives like Don Quixote attacking his windmills. Clearly the giant machines were not rival stags trying to claim their harems. Nevertheless, the whistling sound alone was enough to blind the stags' other senses. All they could see was the image of an enemy which had to be driven away at all costs. The wapiti continued to be killed by the trains until the railroad installed whistles that sounded deeper notes.

There is an old saying that those whom God wishes to destroy, he first makes blind. Both animals and men are endowed with instincts which force them to behave in a certain way and blind them to the consequences. The *Stegomyia aegypti* mosquitoes were blinded by the sympathetic image of a friend, the wapiti by the image of a foe. Thus sometimes their instincts lead living creatures to destruction.

However, to avoid a dangerous conflict between reality and instinct, nature often institutes "control tests" which enable an animal to verify the identity of a friend or enemy. One creature which "tests" the identity of its mate is the grayling butterfly.

We noted that the male *Attacus atlas* moth, which flies at night, pursues a scent emitted by the female. The male grayling butterfly is abroad in the daylight. He sits on a branch or a thistle and waits until he sees a female flying past. He will pursue anything which is (1) large, (2) dark in color, (3) close to him, and (4) fluttering. Of course, butterflies of many species fit this description. For that matter, so do grasshoppers, dragonflies, small birds, falling leaves, and even the grayling's own shadow. The male grayling will pursue all these mimics of a female of his species. However, he subjects the supposed female to three additional tests.

The first test is: Does the object he is pursuing behave like a female grayling butterfly? Females rise into the air, fly around playfully with the male for a moment or two, and then alight. If the unidentified flying object does not behave this way, the male returns to his post and waits for a female to come by.

The second test is a ritual dance. If the female the male is pursuing is willing to alight on a branch with him, the male sits down directly opposite her. Facing her, he repeatedly waves his folded wings up and down over his back. At the same time he extends his antennae to the side and moves them around, tracing a circle in the air. Then he opens his wings and presses the antennae of the female between them, rubbing them with the bright patterns on his wings that look like eyes. Finally he concludes his performance with a bow.

The eyespots on the male's wings contain a fragrance which arouses the female so that she is willing to mate. To human beings, the fragrance smells like tobacco. If a scientist rubs off

Grayling butterfly.

the fragrance from the male grayling's wings with his finger, the female will not respond to the ritual dance. A male with no scent or the wrong scent will be rejected. This scent constitutes the third "recognition test" the male butterfly gives the female before the two can mate.

In part, Nikolaas Tinbergen[1] won the Nobel Prize for his study of the three tests conducted by the grayling butterfly. Thus the scientific world considered the tests a phenomenon of the greatest importance. The grayling's behavior reveals that nature has established additional controls to safeguard animals from the misinterpretation of instinctual signals.

Nevertheless, misinterpretations do occur, for both nature and man can devise mimics. For example, we noted that a carnivorous species of firefly imitates the blinking code of other species to capture prey. Thus an enemy may send a sympathetic signal and then devour the animal that falls into the trap. Moreover, a weak creature may mimic the appearance of a more dangerous animal, thus frightening away its enemies. The harmless hornet moth so closely resembles a hornet that birds never try to eat it.

Animals can perform tests to make sure that they are not deceived by mimics. By the same token, human beings can try to perceive other people more objectively, testing their perceptions to make sure that their instincts have not deceived them. Above all, we should be on guard against those who construct distorted images of other people, making them appear to be our enemies. Such distorted images can result in the extermination of whole races.

We have noted that the eyespots, the "pictures" of eyes on the wings of the grayling butterfly, contain a fragrance which acts as an aphrodisiac. These imitation eyes fulfill another function. As a rule they are virtually invisible, for they are nothing but a blur when the butterfly is in flight; and when it is resting, it usually folds its wings. However, if a bird spies the butterfly sitting quietly on a branch and flies down to eat it, the butterfly instantly opens its wings. The bird is startled to see two large eyes staring up at it. To the bird, the eyes appear to be those of an enemy such as a cat or a marten. Thus the victim, the

butterfly, deceives its enemy into believing that it too is threatened by an enemy. Frightened, the bird flies away, leaving the
butterfly in peace.

A robin literally "sees red" when it sees red. That is, it
always interprets the color red as the red breast of a rival robin.
A robin will attack a bunch of lifeless feathers that have been
painted red. In the same way, a blue jay will peck furiously at
a bunch of blue feathers. A blue jay "sees red" when it sees blue.

During mating season, male stickleback fish attack anything
in their territory that, like them, has a red underbelly. If we
paint red the lower part of a billiard ball or a piece of wood, the
stickleback will attack it. Nikolaas Tinbergen's stickleback,
housed in an aquarium that stood on the windowsill, assumed a
threatening posture whenever a red car drove past on the street
outside.[2]

Many animals have such brief lifespans that they do not
have time to learn what their enemies look like before the danger is at hand. A bird which had to learn from experience what
a cat looks like would be eaten before the lesson was learned. To
survive, most animals must be born with an instinctive knowl-

The iguana responds to an auditory "enemy schema." It knows that
when it hears the cry of a hawk, it is in danger. South American jungle
hunters imitate this cry in order to capture iguanas, which sometimes reach
a length of seven feet. This is almost the only way they can capture an
iguana, for it can take refuge in a tree more than thirty feet tall and then
leap to the ground and scurry away to another hiding-place. Moreover, it
can swim long distances underwater and can bite through nets. However, if
a hunter imitates the cry of a hawk, the iguana will freeze. Absolute im-
mobility may protect it from a hawk, but not from a human being, who can
quickly throw the iguana into a box and cart it away. Thus the iguana's
innate "enemy schema" seals its doom.

At the right is a *Dolomedes fimbriatus*, the largest species of spider in
Germany. It is attacking a crab spider. This class of spiders includes many
species that eat other spiders. It also includes several species in which the
female, which is larger than her mate, eats the male immediately after they
have mated.

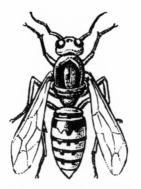

The hornet moth on the right is a harmless moth. Many predators, deceived by its resemblance to the hornet, do not dare to eat it. In other words, they are deceived by their own "enemy schema."

edge of the shapes that spell danger. Their nervous systems are not complex enough to record all the details of an enemy's shape. Thus they respond to abstract symbols.

The fence lizard is a species of lizard native to Mexico and the United States. When the males of this species are about to fight, they put on war paint: blue stripes appear on their sides. One day a zoologist removed the favorite wife from the harem of a fence lizard and painted her belly blue.[3] The harem pasha promptly attacked his wife and drove her away. Then he began to passionately woo a male in his territory which the zoologist had painted gray.

Unfortunately for the fence lizard, the gray intruder immediately attacked him, biting and lashing him with his tail. The pasha was incapable of fighting back. The more savagely he was bitten, the more desperately he tried to court the gray male. Had the zoologist not removed him from the fray, he would probably have been killed.

Signals which trigger instinctive reactions are more powerful than facts.

Even inanimate objects can serve as sympathetic signals and signals of danger. For example, tree snakes and predatory mammals which hunt in trees feel an instinctive dread of branches overhanging water. Of course, they are not consciously aware that an animal hunting over the water might slip and fall in. Instead, they are instinctively afraid when they see a smooth, bright expanse underneath them, and immediately withdraw.

To weaver birds, the signal "smooth, bright expanse underneath me" has a different meaning. The signal does not make them afraid, but gives them a sense of security. Thus weavers like to build colonies of elegant round nests in the branches above rivers. Although they are not consciously aware of the fact, in this location their nests are almost completely safe from predators.

Weaver birds also build their nests over roads, which like rivers conform to the signal "smooth, bright expanse underneath me." Predators too are deceived by the imitation river and leave the birds in peace.

In recent years, some zoologists have begun to speak of "*the* enemy schema." In reality, every animal is born with almost as many enemy schemata as it has enemies.

For example, the pied flycatcher has many enemies, including owls, shrikes, condors, and cats. These enemies do not greatly resemble each other. However, in the same region as the pied flycatcher live some fifty species of birds and mammals which, although they closely resemble the enemies of the flycatcher, will not harm the little bird. Must the pied flycatcher flee from every thrush it sees, simply because the thrush bears some resemblance to the red-backed shrike?

Two signals enable a pied flycatcher to recognize a red-backed shrike: (1) the black stripe extending from the bird's ears over its eyes and forehead, and (2) its general coloring—a pale gray head, throat, and breast, with dark brown wings and tail.

If it sees a wooden model possessing only one of these characteristics, the pied flycatcher is unafraid. However, the bird is terrified at the sight of a wooden bird exhibiting both signals.

Fortunately, the presence of one of the two key signals does not alarm the pied flycatcher. If it did, the flycatcher would spend a great deal of time being afraid. For example, it would fear the thrush, which possesses the same general shape and coloring as the red-backed shrike, and the little ringed plover, which like the shrike has a black band across its forehead.

A different set of signals warns the pied flycatcher when it meets a pygmy owl: (1) a plump, upright form with feathers that range from light to dark, (2) realistic-looking feathers (a carved and painted wooden owl will not frighten the flycatcher, but a stuffed owl will), and (3) symmetrically placed eyes. (The flycatcher pays no attention to a stuffed owl missing one eye.)

Different signals warn the bird against buzzards, hawks, martens, and cats.

The more enemies an animal has, the more complex its "enemy recognition system" must be; and the more complex this system, the more easily the animal can make mistakes or be deceived by mimics. Beyond a certain point, instincts alone do not provide sufficient protection against enemies. At this point, the animal must actually *learn* what its enemies look like.

We have already noted that if a bird had to learn from experience whether or not a cat was an enemy, it would not survive the lesson. Thus to learn, the bird must be taught by other birds that cats are dangerous.

The red-backed shrike, a member of the shrike family, has speared a beetle on a thorn. This bird also captures and eats lizards, mice, and smaller birds like the pied flycatcher.

When songbirds glimpse a screech-owl in the branches of a tree, they begin to chirp and scold. Then all the other songbirds in the area flutter around the owl and join in the song of hatred. This behavior accomplishes two things: It drives away the owl, which has no way of protecting itself against such a storm of abuse; and it teaches the young birds that screech-owls are something to be feared. Birds are born with a vague, instinctive fear of owls. The behavior of older birds reinforces this fear and arms the young birds with a realistic image of their enemy.

Cattle and pigs react with similar hostility to wolves and dogs; caesio fish to moray eels; baboons to snakes; and chimpanzees to leopards.

The psychology of hatred is the same among animals and human beings. Hatred is an instinct triggered by the hatred of other members of the same species. Group hatred teaches an animal what its enemy looks like.

Like group hatred, group flight teaches the young to know their enemies. Reindeer react differently to the barking of different dogs. If they know a dog well, they will ignore its barking and go on quietly grazing. However, if a group of unfamiliar sled dogs passes by, the reindeer will run away.

Actual experience teaches an animal something it can never know by instinct: how to distinguish between "good" and "evil."

In 1913, a young farm boy in the Eifel region of West Germany found a wounded jackdaw on the road and picked it up. He intended to take it home and nurse it back to health. The other jackdaws, misinterpreting his motives, swarmed around screaming their hatred. For the rest of his life the boy was branded as the enemy of jackdaws. Whenever the birds saw him in the fields, they flocked around to scold and scream at him. Sixty years later, the young birds were still being taught to hate him.[4]

Like animals, human beings respond to instinctive "enemy schemata." Western movies and whodunits reveal that humans instinctively distrust whatever appears black, foreign, threatening, dirty, inferior, brutal, or treacherous. Filmmakers capitalize on our instinctive reactions to certain kinds of symbols. The spectator, whether he is American, Japanese, Eskimo, European, or an African from the bush, immediately knows that "that man there is a bad guy," even before the villain has perpetrated any wrongs.

Moreover, like animals, human beings respond emotionally to what they believe to be the image of an enemy. Then, to justify their feelings, they engage in expressions of group hatred and fight together against a common foe. Feelings of fear and envy feed their hatred. The government, the press, radio and television may all disseminate propaganda to intensify the hatred of the people for their alleged enemies. Writers and other propagandists aid in constructing an enemy schema. The schema is basically the same whether the Nazis are attacking the Jews, the whites the blacks, the Irish Protestants the Catholics, or the Establishment political activists. The enemy is always portrayed as dangerous and inferior. Responding to this schema, people move from words to deeds and begin to exterminate other human beings as if they were some species of insect. Meanwhile, those who constructed the enemy schema pretend to be innocent of the horror that their own words have brought about.

The only way to combat this sort of mass hysteria is to become more aware of how enemy schemata affect the unconscious mind. Our instincts can make us the playthings of demagogues. Human beings tend to feel superior to the wapiti which attack whistling locomotives, or the robin  which pecks at a clump of red feathers. But human beings can be as readily deceived by their instincts as any animal. We may not respond to elementary signals like a whistle or a color. Instead, we respond to words, which paint a black-and-white picture of an enemy. Soon we find ourselves believing things simply because we want to believe them. Thus we must all be on our guard against instinctive reactions.

I have often spoken of instincts, reflexes, and instinctive reactions. Exactly what are instincts and reflexes, and what role do they play in courtship and mating? To clarify these terms, I will devote the next two chapters to the concepts of reflex and instinct.

# PART THREE

## Irrational Forces

# CHAPTER 5

# Brainwashing and the Perversion of Emotion

*The Conditioned Reflex*

A shrill alarm clock went off and five dogs went to lie down in the corner. In a moment they were sound asleep. What caused this paradoxical behavior?

Carmine Clemente had inserted a thin wire through the cranium of each dog and into the sleep control center of the brain, which is no larger than one cubic millimeter in size.[1] To state it simply, when electrical nerve impulses signal the sleep center that the body is tired, the nerve cells of the brain send out electrical signals that put the body to sleep. In Clemente's experiment, the wire inserted into each dog's sleep center acted as an artificial nerve which signalled the brain that the dog was tired. Thus the animal fell asleep at once.

Having completed this phase of the experiment, the scientist began to ring an alarm clock shortly before putting the dogs to sleep. After repeating the process twenty times in three days, he could make the animals fall asleep simply by ringing the alarm clock. Whenever the alarm went off, nerves in the ear signalled the brain to put the dog to sleep. These nerves had taken over the job of the wire, the artificial nerve, which could now be removed. This kind of reaction is called a conditioned reflex.

Both animals and human beings can be manipulated by means of conditioned reflexes, which are adaptations of already-existing reflexes. Exactly what is a reflex?

Erwin Lausch describes the phenomenon as follows: "We function by means of innumerable reflexes. When a speck of dust gets into one of our eyes, we immediately begin to blink, and tears flow to wash the dust away. When we walk from a dark room into brilliant sunshine, the pupils of our eyes contract. If something gets into our windpipe, we gag and cough until it is forced out. When the rectum is full, the sphincter muscle contracts. There are swallowing and choking reflexes, reflexes to regulate breathing and circulation, stomach and intestinal activity, the production of saliva, bile, and gastric juices. It has been estimated that more than twenty thousand reflexes protect the human organism and enable it to function as harmoniously as possible, sparing the human brain the effort of making constant decisions." [2]

The reflex is an involuntary nervous mechanism. Sensory cells receive a stimulus and signal the central nervous system, where the impulse is transferred through intermediate cells to other nerves. These nerves signal different parts of the body to react in an appropriate manner. Reflexes are automatic and unconscious, like the knee-jerk reflex activated when a doctor taps us on the knee. Unlike instinctive reactions, reflexes do not involve the emotions. They are as mindless and unpremeditated as the actions of a robot.

Sometimes when we talk on the telephone, we may hear distant voices talking on another line. The same kind of cross-circuiting can interfere with the normal functioning of a reflex. This happens when a second, hitherto meaningless signal occurs in association with the stimulus that produces the reflex. Involuntarily an animal or human being learns to associate the new signal with the original stimulus, so that the signal becomes a so-called "conditioned stimulus."

The best-known example of this phenomenon is Ivan Pavlov's famous experiment on dogs. A dog begins to salivate when it sees food. This is a natural or unconditioned reflex. But if a bell is rung shortly before the animal receives its food, after a time it will begin to salivate whenever it hears the bell, even if it no longer receives any food. When an animal is trained so that it responds to a hitherto meaningless signal in the way that it formerly responded to a natural stimulus, the resultant reflex is called a "conditioned reflex."

The phrase may sound harmless enough, but it conjures up frightening visions of our power to manipulate human behavior. There is no surer way to alter behavior than to train human beings to obey conditioned reflexes. Psychiatrists engaged in behavioral therapy may use this technique to benefit people with problems; others may employ the technique in brainwashing.

John B. Watson, the founder of behaviorism, offers an example of how the power to manipulate behavior can be abused.[3] Watson used to sound a gong before he entered the nursery to scold his eleven-month-old son Albert. After a time, the child became frightened whenever he heard the sound of the gong. This kind of treatment can cause severe emotional damage in later life.

Once Watson had established the conditioned stimulus, he could manipulate his son as he chose. The boy enjoyed playing with guinea pigs, but his father did not wish him to do so. Whenever Albert crawled over to the animals, Watson struck the gong without saying a word. After a time, the boy was so afraid of guinea pigs that he ran away crying whenever he saw the animals, even if they were a long distance away. Moreover, he grew afraid of other furry things such as rabbits, fur coats, and even the beard of a Santa Claus mask.

A person who has been trained to feel fear can be cured of his fear by retraining. John B. Watson began the cure of his son by showing the boy guinea pigs from a long distance away. At the same time he gave his son chocolate. Every day he repeated the experiment three times, bringing the animals nearer and nearer to the boy. Finally Albert got over being afraid.

Unfortunately, when a doctor is trying to cure patients of

neurotic fears—fears they have been trained into feeling—chocolate is simply not enough.

In the 1960's a school of psychiatry known as behavioral therapy came into existence. Therapists of this school attempt to cure abnormal behavior patterns by much the same methods as Watson used to cure his son. They have been criticized because their methods closely resemble the techniques of brainwashing. They use a wide-spectrum methodology that includes the administering of electrical shocks. A cure is effected when the patient is made to feel aversion to "undesirable" habits or behavior patterns, such as homosexuality, fetishism, other forms of sexual deviation, alcoholism, delusions of persecution, or agoraphobia.

For example, a homosexual might wish to be cured of his homosexual orientation. In the past, he has always experienced a pleasant sensation when he looked at certain kinds of films and pictures, even though he was aware of the "abnormality" of this sensation. When he enters into "aversion therapy," the patient is attached to a machine which records his emotional reactions. Then the therapist shows him the same kind of films and pictures that have always given him pleasure. When the machine shows that the patient is experiencing pleasure, he receives an electric shock. After many such treatments, he develops an aversion to any kind of homosexual reaction.

Psychoanalysts criticize aversion therapy on the grounds that it treats the symptoms of a disorder but not the underlying cause. Moreover, this mode of treatment does bear a striking resemblance to brainwashing, except that the patient voluntarily chooses to undergo it.

Animals are helpless to resist training through electrical shocks or electrical stimulation of the brain. In America in 1970, a scientist employed this barbarous technique to train an aggressive chimpanzee.[4] With the aid of a computer, he taught the animal to be gentle.

Like the dogs which fell asleep at the sound of an alarm clock, a chimpanzee named Paddy was trained by means of two wires inserted into his brain. One wire was introduced into the aggression center of the brain. Every time the ape grew angry, the wire in the aggression center signalled that he was about to suffer a fit of rage. This information was tapped from the wire and fed into the computer. The computer then sent electrical signals along the second wire into the area of the brain that registered aversion. Thus the chimpanzee began to associate anger with aversion. Now that he no longer enjoyed his fits of rage, the ape grew gentle as a lamb. Such an exchange or reversal of emotions is a special form of conditioned reflex.

Apart from man's initial intervention, the entire training process took place within the chimpanzee's own nervous system. A normal chimpanzee experiences a reflex, a sense of satisfaction, when it is allowed to vent its anger. Electrical stimulation of the brain created a conditioned reflex, associating anger with aversion rather than satisfaction.

By using the same methods on human beings, we could associate the sexual act with feelings of disgust, thoughts of suicide with pleasure, eating with shame, and stealing with a sense of justice. In other words, we could duplicate in normal people sensations experienced only by perverted, suicidal, paranoid, or criminal personalities.

Moreover, to produce such effects we would not even need to resort to electroshock or electrical stimulation of the brain. Behavioral scientists have developed far subtler methods. In police states, political prisoners are frequently placed in solitary confinement, kept in perpetual darkness or a brilliantly lighted room, and forced to stand up all day or to listen to monotonous noises. Their sleep is continually interrupted. They may be told that they will shortly be executed. At night they learn to dread the next day, and during the day they dread the night. When a prisoner's resistance has been worn down, he is given a little hope: If he will simply agree to a few of the more acceptable theories or policies of the present government, his treatment will improve. The prisoner is rewarded for every gesture of cooperation. Soon a conditioned reflex is established. Whenever the prisoner thinks of his past beliefs, he associates them with pain so that they appear to be destroying him. Meanwhile, adaptation to the present regime becomes associated with the idea of survival. This method is more time-consuming than treatment with electrical shocks, and the prisoner's wardens must be skilled in the arts of manipulation. On the surface it appears more human than electroshock treatment; but in reality brainwashing is even crueller than the physical tortures of medieval times, for there is no way to resist it.

Brainwashing can completely reverse all a person's normal emotional reactions, turning fear into pleasure and affection into loathing. It can make a man betray everything that was formerly dear to him. Brainwashing represents a form of mental murder.

Animals which have undergone brainwashing have an easier time of it. Two weeks after the last electrical charges were sent into his brain, Paddy the chimpanzee began to get angry again. Moreover, human patients treated with electrical shocks to cure them of homosexuality have reverted to their former behavior patterns after a period of several months or years. However, former inmates of concentration camps, whom the Nazis brainwashed into helping to execute their fellow-prisoners in the gas chambers, have remained emotional wrecks for the remainder of their lives. The same is true of many Americans brainwashed by the North Koreans.

Perhaps brainwashing techniques would never have been developed, had it not been for Ivan Pavlov's discovery of the conditioned reflex. In any case, clearly the science of animal behaviorism has profoundly affected all our lives. And like atomic physics, it is Janus-faced: It can harm living creatures as well as help them.

The advertising industry makes extensive use of brainwashing techniques. A German magazine containing eighty-six ads was published in May, 1973.[5] Only ten of the advertisements contained any specific information about the product. Seventy-six made a purely emotional appeal to potential buyers. Remarkably, only three products—suntan lotion, bath oil, and wine —used "sex appeal" to attract buyers.

Above all, the advertisements appealed to the consumer's vanity, his desire for prestige. Twenty-seven of these modern fairy tales claimed that the product in question—champagne, beer, lemonade, milk, deodorant, perfume, razor blades, mortgage bonds, synthetic fabrics, cigarettes, cars, and lipstick—was used by the aristocracy. Even the army appealed to the public's desire for prestige. Eleven advertisements promised people a more comfortable life. Nine created anxiety so that the product could appear as a savior in distress. The latter group included cars, insurance, margarine, insect spray, toothpaste, and analgesic tablets. One ad for fruit-flavored chewing gum appealed to the desire to conform: "What millions do, you should do too." Four ads, for motorbikes, portable radios, film cameras, and beer made just the opposite appeal: "You'll be the only one." An automobile firm tried to frighten its clientele: "If you wait too long, they may all be gone." Producers of wine, mineral water, venetian blinds, and electricity tried to convince the public that buying their product amounted to "making the right decision." Firms selling skin cream, beer, fruit juice, and film cameras wanted to bring people closer to nature. Five other firms imitated popular quiz shows.

In several cases, the advertisements attempted to reverse normal emotional responses to a product. For example, the ads identified milk with masculinity, beer with exclusivity, lemonade with the British nobility, chemical skin creams with naturalness, frozen foods with hunting, razor blades with comfort, and mortgages with earning money. (At this time, the stock market was plunging.)

This kind of mass manipulation is alarming because it shows how easy it is to control public opinion. Millions of otherwise quite sensible women go through life worrying about their laundry because television commercials associate a clean wash with having a clean conscience.

The same kind of mass manipulation, employed by governments and politicians, can drive nations into a homicidal or suicidal frenzy. During World War II, propaganda exploited the feelings of the Japanese regarding the Samurai virtues—masculinity, honor, service, the defense of one's country, self-denial. The result was the suicidal missions of the kamikaze pilots, who sacrificed their own lives to destroy an enemy ship.

By training animals and human beings to obey conditioned reflexes, we can radically alter behavior patterns. Pavlov's experiments were limited to making a dog salivate when it heard the sound of a bell. These experiments have led to others, so that now it is possible to change basic emotional responses. Moreover, conditioned reflexes can trigger the development of additional behavior patterns which do not themselves constitute conditioned reflexes. Individual human beings may be manipulated by behavioral therapy, brainwashing, and advertising, and masses of people may be programmed to kill or to commit suicide.

Human beings must exercise their free will to resist the forces of manipulation. For example, advertisements make irrational appeals to the emotions. Consumers should resent these appeals and make every effort to examine products critically before buying. However, under certain conditions—for example, when a man is being brainwashed—he has very limited powers of resistance. He can no more resist manipulation than a person can keep his knee from jerking when it is tapped.

Reflexes and conditioned reflexes are not the proper subject of this book. Thus I will briefly summarize several points before turning to the question of instinct.

1. In addition to reflexes which produce a certain kind of behavior, there are also reflexes which inhibit or repress behavior.

2. In addition to conditioned reflexes, there are secondary conditioned reflexes.

For example, let us assume that a dog has already been trained to obey one conditioned reflex. Instead of salivating when it sees its bowl of food, it salivates when it hears the sound of a bell. Then shortly before the bell rings, a signal light is turned on. After a while the animal will begin to salivate when it sees the light.

Political demagogues often make use of secondary conditioned reflexes to manipulate human beings. They may even exploit additional reflexes created by association with secondary reflexes, i.e., reflexes at third or fourth remove. If a demagogue wishes to radically alter public opinion and knows that he cannot do so in a single step, he will use "salami tactics," carrying out his plan step by step or reflex by reflex, the way one cuts off slices of salami.

3. At times a demagogue may find that he cannot institute the conditioned reflexes necessary to control human behavior. In this case, he may attempt to produce conflict by establishing two contradictory conditioned reflexes. Then later he can step in and capitalize on the conflict.

For example, let us assume that a dog has been trained to enter the kennel to get food whenever it hears a bell ring. Once this reflex is established, the dog is taught never to enter the kennel when a lamp on the roof is turned on. If it enters the kennel when the lamp is on, it is punished with an electric shock. When the dog has been trained to obey both reflexes, the two signals are issued simultaneously. The dog wants to run into the kennel to eat its food. At the same time, it is afraid of receiving an electric shock. As a result, the animal is confused. It runs back and forth, circles around the kennel, howls, and tucks its tail between its legs. Torn by ambivalent feelings, it does not know what to do. Now it can easily be taught to turn on its kennel-mates and bite them for no reason at all.

The science of animal behaviorism can readily be exploited to manipulate human beings. I believe that fewer than one in ten people have the critical perception necessary to detect when their emotions are being manipulated for unscrupulous ends; and fewer than one in twenty have the inner strength to resist this manipulation. Perhaps this book may add a few to the number.

# CHAPTER 6

## Instinct, the Motive Force Behind Behavior

*What Is an Instinct?*

When a man feels hungry, he goes to a grocery store, buys food, takes it home, cooks it, and eats it. He would not buy food just for the fun of it and then throw it away.

Cats, however, frequently hunt not because they are hungry, but "just for the fun of it." If they did not enjoy hunting for its own sake, they would not survive. A cat is not intelligent enough to reason that when it is hungry, it must go to sit beside a mouse hole and wait until a mouse comes out so that it can kill and eat it. Instead, cats enjoy each step involved in hunting a mouse quite separately and for its own sake. Often a cat will simply pretend to lie in ambush and stalk prey. It will sit watching the drain in an empty bathtub, not because it expects to see a mouse emerge, but because it enjoys the game of lying in wait. Moreover, cats enjoy capturing things, even if all they capture is a ball of yarn, a rag, or another cat's tail.

Four separate instincts impel a cat to stalk, capture, kill, and eat its prey. Some of these instincts are stronger than others. To discover which instinct was weakest, i.e., most quickly appeased, zoologist Paul Leyhausen supplied cats with a steady stream of mice. First the cats stopped eating their prey. But even after their physical hunger was appeased, they continued to kill the mice. Then they tired of killing, but went on stalking and capturing the mice. Eventually they grew bored with capturing. Last of all they gave up stalking.

This hierarchy of instincts enables the cat to function efficiently. A cat must stalk its prey for a long time before it has a chance to capture it. Thus the stalking instinct must be stronger than the instinct to capture. Moreover, a captured mouse can easily escape, so the cat must be more intent on capturing than killing it. Similarly, a cat may be driven away from its prey, so the instinct to kill must be stronger than the instinct to devour.

In passing it should be noted that even in predators, the instinct to kill is not as strong as many people suppose.

The hierarchy of instincts in the cat explains the apparent cruelty of cats to their prey. A house cat which has just finished its dinner may have satisfied its hunger, but not its desire to stalk, capture, and kill. Thus leaving its food bowl, it goes straight to the cellar to hunt for mice, and plays tag with the mouse before killing it and leaving it on the ground. The cat does not care whether it is playing with a mouse or a ball of yarn. Its behavior is not sadistic or cruel, for it has no conception of the torture suffered by its victim. Only human beings are capable of enjoying the torture of another living creature. The cat simply enjoys playing a game.

Each of the cat's four autonomous instincts—stalking, capturing, killing and eating—functions only to the degree demanded by the animal's environment. As a rule, wild jungle cats do not play with their prey.

The behavior of cats reveals the following facts about instinctual behavior in general:

1. Many people mistakenly interpret the hunting behavior of cats in terms of a single instinct, that of hunger. In reality, instinctual behavior patterns frequently consist of a series of

individual, autonomous acts, each motivated by a separate instinct. For example, the "sexual instinct" is expressed in a series of separate acts—the search for a partner, the recognition of the partner, the overcoming of aggression and of the instinct to flee, the synchronization of the two partners' readiness to mate, and the actual act of mating.

2. Instinctual behavior differs radically from the reflex. We might say that instinct represents a higher order of reflex. Behavioral psychologists interpret all animal and human behavior as resulting from conditioned reflexes. However, at times they make the mistake of confusing instincts with reflexes.

We have already noted that a relatively simple neural mechanism underlies each reflex. The reflex arc consists of a nerve which transmits an impulse to the central nervous system and through adjustor neurons to another nerve which orders the glands or muscles how to behave. The reflex is involuntary, like the actions of a robot.

When an instinct is involved, the processes taking place in the central nervous system are more complex. Above all, hormones affect the adjustor neurons, intensifying, inhibiting, and varying their responses much as coffee, sleeping tablets, narcotics, and other drugs influence our brains.

The point at which the nervous impulse passes from one neuron to the next is called a synapse. Hormone A may accelerate the transmission of neural signals in one group of synapses, inhibit it in another, and have no effect at all in a third. Hormone B may have the same effect in other synapses.

Hormones are manufactured and stored in the body. At times hormonal supplies are depleted, and it take time to build them up again. This means that unlike a reflex, instinctual be-

One hypothesis suggests that the loving relationship between mothers and children is the source of all tender exchanges between male and female. Female squirrels are very loving mothers and spend months caring for their young. The female squirrel is not very affectionate toward the male. However, shortly before mating and for several days thereafter, the male behaves like a baby squirrel. During this period, the female shows the male the same affection she shows her young.

havior is variable. The inclination to obey an instinct varies with the quantity of hormones affecting the nervous system at a given time.

It is basic to the theory of instinct that the readiness to obey an instinct partially depends on "motivation," on an inner compulsion or drive to behave in a certain way. A reflex is instantaneous and bypasses the conscious mind, However, before carrying out an instinctive act, we consciously feel the impulse to do so. When an impulse becomes conscious, we may experience a conflict between instinct and reason. As a rule, reason becomes the tool of instinct. We know little about our instincts, for we all suffer from the illusion that we are their masters rather than their slaves.

An instinctual act is one which springs from a sense of inner necessity.

When a cat that loves to hunt spends weeks tamely eating from its bowl and does not hunt a single mouse, the drive to capture prey builds up and grows very strong. Soon it is so strong that the cat will use anything—a ball of yarn, houseflies, or its own shadow—as a substitute mouse. Eventually the drive from within becomes so urgent that no external stimulus is required to trigger the instinct. At this point the cat may begin to stalk invisible prey.

Thus the longer a man or animal fails to carry out an instinctual act, the stronger grows the drive to perform it. Moreover, the stronger the drive becomes, the lower is the threshold of the stimulus that finally triggers the act.

The first step in the performance of an instinctual act is the reception of a stimulus. That is, the sense organs perceive something that triggers the act. Of the many impressions received by the senses, only a few are passed along the nerves to the "action center" of the brain. These few impressions correspond to symbolic signals designed to trigger a specific instinct. Sometimes the sense organs themselves "filter out" irrelevant signals. For example, the "ear" of the male *Stegomyia aegypti* mosquito (the species that carries yellow fever) simply does not hear notes higher or lower than the humming note produced by females of its species.

The antennae of the male *Attacus atlas* moth smell various fragrances, but only a few special sensory cells respond to the sexual attractant produced by the female. When these cells receive the appropriate signal, the nerves convey the signal directly to the neural nucleus that compels the moth to fly toward the fragrance.

Whether the *Stegomyia aegypti* mosquito and the *Attacus atlas* moth obey a reflex or an instinct is a debatable point. However, there can be no doubt that the herring gull acts on instinct. When the gull chooses a mate, its brain filters out signals which are not "sympathetic." Only if the bird sees that the eyes and eye rings of its potential mate are the right color will it proceed to the next step, actual courtship.

Thus a stimulus must "fit" into the nervous system as a key fits into a lock. The "lock," or neural mechanism, into which the stimulus must fit is called an innate releasing mechanism, or IRM, and the key signal itself is called a key stimulus.

Among human beings, group W facial features are key stimuli of the IRM which causes group C types to feel attraction.

The innate releasing mechanism is not the only factor which determines an animal's readiness to obey an instinct. The animal's threshold, i.e., the strength of the stimulus required to trigger an instinct, continually varies. For example, when a shark first attacks a school of mackerel, it moves slowly and calmly. But after it has devoured its first victims and the water is turning red with mackerel blood, the shark's threshold is lower and it embarks on an orgy of killing.

Thus when an animal is repeatedly stimulated to perform an instinctual act, its threshold becomes lower and the key stimuli required to trigger the act may be correspondingly weaker. An animal which has repeatedly performed a certain act and which consequently performs it more readily than before is said to be sensitized.

If an instinctual act is not performed for a long period, the instinct can atrophy. When cats grow fat and lazy, they cease to be interested in chasing mice. Moreover, if any animal performs an instinctual act so often that it becomes a habit, the animal may cease to enjoy the act, and once again the instinct can atrophy.

On the other hand, the repetition of an instinctual act may give an animal increasing pleasure. As the instinct is "trained," the animal becomes more and more inclined to obey it. Unfortunately, aggression often functions in this manner. In his book *On Aggression,* Konrad Lorenz shows that when an animal or man enjoys the performance of an aggressive act, his appetite for aggression grows. Some people mistakenly believe that it is good for them to express their aggression. In reality, they are training their aggressive drive and consequently becoming more aggressive.

Many factors affect the instinctual behavior of animals and human beings. An animal's instincts mature along with its body. As nerve fibers grow, neurons are linked by synapses. The body begins to produce various hormones. External forces can encourage or inhibit the process of maturation. Later in this book I will discuss females which feel no mating drive until males have spent weeks courting them.

Unlike reflexes, instincts are not fixed and immutable forces. An animal's experience can modify its instincts. For example, we have seen that jackdaws can learn what an enemy looks like and communicate this knowledge to other jackdaws. Moreover, as an animal matures, it passes through phases when it is highly susceptible to environmental influences. In conjunction with heredity, the animal's personal experience helps to mold its personality. I will discuss this subject more extensively in my future book concerning the relations between parents and children.

The readiness of an animal to obey an instinctual impulse fluctuates with the time of day. For example, at night human beings are more timid and easily frightened than during the daytime; and in daylight they are more aggressive than at night.

The strength of instinctual drives also varies with the seasons. Many animals feel the urge to mate for only a few weeks, days, or hours of the year. For the rest of the year, they are completely uninterested in sexual relations. Hormonal activity governs the mating cycle.

We might describe an instinct as an inherited drive to behave in a certain way. The drive is regulated by hormones and by the activity of the nervous system. As a rule, instinctual behavior is triggered by an external stimulus. However, at times it may be triggered spontaneously from within. When obeying a reflex, the animal responds to the stimulus by immediately performing the required act. In the case of an instinct, the response is not an act but an emotion: a feeling of anticipation if the animal is about to perform a pleasurable act, a feeling of aversion if it must avoid a threat. The emotion drives the animal to perform the required act. The reward for having done so is a feeling of satisfaction.

We have already noted that if a long time has elapsed since an animal last performed an instinctual act, no external stimulus is required to trigger the act. In this case, spontaneous emotion compels the animal to seek an external stimulus for the instinct. When a cat stalks through the tall grass of a meadow, it is seeking an external stimulus for its desire to capture and kill. That is, it is looking for prey. When an animal searches for a stimulus to trigger an instinct, its behavior is called appetitive or searching behavior.

A man whose threshold of aggression is momentarily very low, will go around looking for someone with whom he can pick a quarrel. Similarly, a wild rabbit that has spent some time vainly searching for a mate will eventually become so desperate that it will indiscriminately woo any female it sees.

When a cat hunts mice, it performs several separate activities, each governed by a separate instinct. Each activity is a form of appetitive behavior, a search for a stimulus which will trigger the next phase of the hunting process. The cat stalks in order to capture, captures in order to kill, and kills in order to eat.

When an animal obeys an instinct, an emotion precedes, accompanies, and follows the act. A man feels angry before he begins to quarrel. During the quarrel, he experiences a wide gamut of emotions. For example, he may feel afraid if the other person attempts to fight back. When the quarrel is over, he may feel triumphant if he has won the quarrel, depressed if he has lost.

When our instincts come into play, we experience at least one of a wide spectrum of possible emotions. We feel a somewhat different emotion if we are anticipating something good to eat, a love affair, a journey, a social coup, a game, or a sum of money. Hormones, good and bad memories, hopes and fears and inhibitions play a role in how we feel; and our feelings in a given situation change in accordance with how other people react to us.

Animals and human beings possess as many instincts as they do emotions and shades of emotion. Whenever we experience an emotion, we know that our instincts are involved.

Some years ago there existed a school of zoologists strongly opposed to anthropomorphism, to the attribution of human insight, morality, and emotions to lower animals. Television animals like Lassie, Skippy the kangaroo, Flipper the dolphin, and the pets of "Daktari" are products of anthropomorphism. No real animal thinks or feels like these fictional creations. On the other hand, it would be wrong to refuse to attribute to animals any emotions remotely similar to ours. If a dog wags its tail when it sees its master, some very scrupulous zoologists would simply describe the tail-wagging as a "mechanical reaction." I believe that in such a case it is quite permissible to say that when the dog wags its tail, it is happy to see its master.

The refusal to attribute emotions to animals is a holdover from the days when animal behavior was interpreted solely in terms of reflexes. "Who knows whether the animal feels anything when it performs an instinctual act?" zoologists said.

At one time people believed that only human beings experience true emotions, for only human beings are capable of expressing emotion in music and poetry. In reality, the difference between animals and human beings lies in man's superior intelligence, ability to speak, and ability to transmit knowledge to his descendants. The higher animals have just as many emotions as we do. In fact, many of the higher animals experience a more intense emotional life than human beings. Our intellect and ability to speak enable us to express our emotions in artistic form. However, electrocardiograms and electroencephalograms show that the higher animals experience stronger and more abrupt surges of emotion than human beings. On the other hand, these surges of emotion die down more quickly than in humans.

Only the higher animals experience emotion. The behavior of animals like protozoans, sponges, and jellyfish is regulated solely by automatic reflexes. To experience emotion, a species must have developed instincts as well as reflexes. When a species develops instincts, emotions become the motive force behind behavior. For example, human beings undertake few actions unless they feel driven or compelled to do so. Virtually everything a man does, he does in response to various drives: the need for security and prestige; the social bonding instinct; fear; and the sexual and aggressive drives.

Religious zeal may be regarded as an attempt to guarantee our security in a future life. Thus it too springs from an instinctual drive. Idealism and altruism derive from the same source, in conjunction with the social bonding instinct. Performing a kind act gives us a feeling of satisfaction, and in anticipation of experiencing this feeling, we are eager to help other human beings. Unrecognized geniuses too can derive so much satisfaction from their work that they can live without recognition and approval from others.

Frequently we deceive ourselves and each other about the real reasons for our actions; but the true cause is always emotional, i.e., instinctual in nature.

The will plays a pivotal role in human life, for it mediates between the world of instinct and that of reason. On the one hand, the will is a rational force which we consciously employ to make ourselves perform or refrain from performing certain acts. On the other hand, the will itself resembles an instinct.

A human being is born with a strong or weak will. Training and experience can further strengthen or weaken the will. Moral values or a sense of responsibility for our acts may cause us to exercise our will to modify our behavior. However, we do not always have the strength to act as we choose. Our will gives us the freedom to choose among various alternatives; but freedom of the will is always relative, not absolute.

The will mediates between instinct and reason. Neither the one nor the other, it has something in common with both.

What would a man be like who consisted of pure intellect,

without any emotions? He would have no children, for to him children would simply represent a lot of trouble. He would have no use for love. He would not listen to music or look at paintings, for no kind of art could make him feel anything. He would be indifferent to food and to life itself. Such a person would resemble a robot. He would have no reason to do anything or even to keep himself alive. To function, a computer needs someone to program it, to turn it on and feed it with data. Without someone to operate it, it is simply a metal box without initiative.

In recent years, many human beings have destroyed their emotional and instinctual life by the use of narcotics. They have turned away from concrete goals and real satisfactions to live in a world of dreams. Their desire for gratification is effortlessly satisfied by the drug. Thus they need not seek pleasure in the realm of deeds and facts. The "trip" destroys their drives. As a result they cease to function, becoming like computers that no one has turned on.

Not the achievement of a goal, but the performance of an instinctual act, satisfies an instinct. A cat may capture a mouse the first time she leaps at it. She has now achieved her goal, but her drive to capture prey is far from satisfied. Thus she continues to play with her prey. When she has satisfied the drive to capture, she may experience the drive to kill.

When a golden hamster feels the urge to run, it runs—outside if it is living on the Asian steppes, or on its little treadmill if it is locked inside a cage. Running on a treadmill seems to us a futile occupation; yet the hamster will go on running all day and is acutely unhappy if anyone takes the treadmill away.

We are all governed by drives much like that which causes the hamster to run on a treadmill. That is, our drives are satisfied by the performance of an instinctual act, not by the achievement of a goal. An aggressive man who is looking for a quarrel will not be satisfied if his opponent tries to settle the quarrel amicably by acceding to his demands. In reality, the aggressive man wants to go on quarrelling. The more others give in to him, the angrier he will become until he gets the battle he wants. He is not trying to achieve a specific goal, but desires a fight for its own sake. Only by quarrelling can he release his pent-up aggression.

Thus before we quarrel with anyone, it is important to know the real cause of the quarrel.

Let us summarize what we know about instinctual behavior. Instinctual behavior consists of a series of segments of behavior, each of which is triggered by a separate, autonomous instinct. Instincts vary in strength. The strength of an instinct is regulated to give the animal the best chance of survival in its natural environment.

Physically, instinctual behavior is linked to a mechanism consisting of neurons and hormones which affect the neurons. Because instinctual behavior is linked to a physical mechanism, instincts mature with the body of the individual. Moreover, hormones cause instincts to vary in strength at different times of the day and year.

The strength of the stimulus required to trigger an instinct varies with the length of time that has elapsed since the last time a given instinctual act was performed. When an instinct is not released for a prolonged period, the animal or human will seek a stimulus to trigger the instinct. The stimulus may be a symbolic signal which conveys a message to the innate releasing mechanism; or it may be a more complex image that the animal must learn to recognize. An object bearing little resemblance to the key stimulus may also trigger an instinctual reaction. (For example, a ball of yarn may serve as a substitute mouse, thus triggering a cat's instinct to capture.) Lacking a substitute for the key stimulus, the animal may spontaneously perform an instinctual act. (I.e., the cat may stalk an invisible mouse.)

Animals can be sensitized to instincts. If an instinct is not released, or is released so frequently that it becomes a habit, it can atrophy. However, repeated performance of an instinctual act can heighten the animal's pleasure and thus its inclination to perform the act. Disappointment or punishment inhibit the performance of the act.

Instincts are not fixed and immutable, but can be modified by experience. Early experiences, willpower, personal morality, and the sense of responsibility can intensify, alter, inhibit, or block an instinctual act, especially in human beings. As a rule, we are unaware of the true motives of our behavior and thus fail to recognize that we are acting instinctively.

After perceiving the key stimulus and before performing an instinctual act, an animal experiences an emotion which compels it to act or inhibits it from acting. Emotions play no role in reflexive behavior. However, by training an animal to obey a conditioned reflex, we can alter its emotional reactions. The reaction to a conditioned stimulus is neither instinctual nor reflexive. Nevertheless, the use of the conditioned reflex enables human beings to duplicate emotional reactions normally associated with instinctual behavior.

Not the achievement of a goal, but the performance of an instinctual act, satisfies an instinct.

It is difficult to define the exact nature of an instinct. We can build an automaton which exactly simulates a reflex. Moreover, a reflex can be precisely defined as a chain of electrical impulses. Instincts, on the other hand, involve emotion. Thus an instinct is a far more complex phenomenon than a reflex.

For example, we can construct an electronic cat, a metal figure on wheels, run by electric motors and guided by an automatic steering system, with photoelectric cells for eyes. We can program this mechanical cat so that it will stalk prey, lie in ambush outside mouse holes, and capture and swallow mice. We can program the cat to respond to external stimuli. However, we cannot program it to feel the pleasure of the chase, fear, or satisfaction in the kill.

What is an emotion? Scientists hypothesize that emotions are linked to a combined neural-hormonal mechanism. However, we do not have the slightest idea *how* such a mechanism actually produces the emotions—urges or inhibitions—associated with instinct. In fact, we do not even know whether emotions

During mating season in May, the male frog called *Rana esculenta* is aggressive and brutal. The males seize the females like wrestlers and squeeze them to force out the eggs. In the great marriage market of the frog pond, the male indiscriminately grabs hold of anything that is the size of a frog. He will even grasp another male frog, a toad, a dead frog, or a piece of wood. The frog in the picture cannot capture his own image in the mirror, and so has adopted an imposing and aggressive posture.

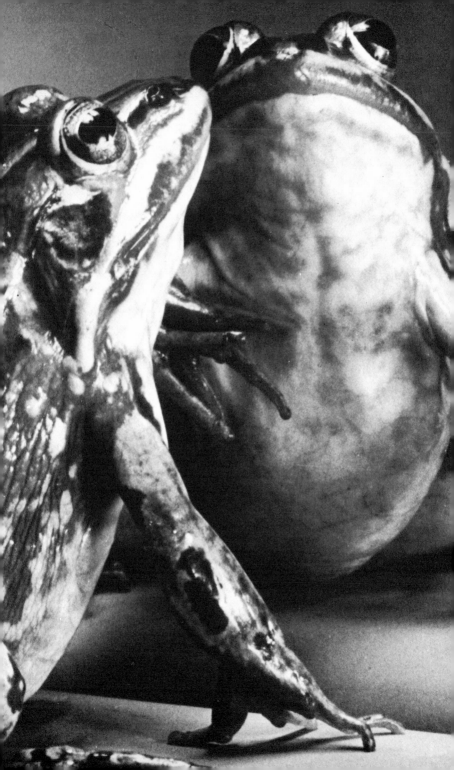

are the product of a physical mechanism which can be analyzed and duplicated, or whether they result from a phenomenon incomprehensible to human beings. As long as we do not know what produces emotion, the nature of instinct will remain a mystery. To be sure, we can investigate instinctual behavior and draw practical conclusions regarding the psychology of animals and human beings. But we can no more define what an instinct really is than we can define matter, energy, space, or time.

Some people feel that instincts are by nature something base, primitive, animalistic, and evil. In reality, instincts are no more evil than emotions are. Some emotions lead to sublime, others to base acts. If it were not for the instinct to band together in groups and to form lasting ties with a mate, human beings would be ill-tempered and solitary creatures like the polar bear.

All noble and ignoble ideals influence our behavior by stirring our emotions. In the final analysis, we cannot rationally justify our belief that there are certain things that a decent person must or must not do. Instead, such beliefs are based on a feeling of inner necessity. That is, they are based on innate social instincts. These instincts form the foundation of character. If children are not raised properly, the foundation can be destroyed. But if our basic instincts are not perverted, they are a more reliable guide to behavior than all the moral commandments and sociological theories in the world. These social instincts, which defy rational analysis, serve as a sort of "divine voice" inside us. In fact, they form the basis of conscience.

To be sure, the realm of instinct and emotion is not only the seat of conscience. It is also the source of the demonic, that overwhelming force that unstable and destructive men either cannot or will not control. Almost without exception, men who have power over others succumb to demonic forces. Power and the abuse of power go together like thunder and lightning.

The demonic is the authentically evil. It is a yielding to one's worst instincts, to the point of harming other human beings and consciously suppressing the counsels of reason and morality.

We should draw a clear distinction between the concept of sin and that of evil. Sin implies the yielding to instincts in violation of religious commandments, without harming other human beings.

In the past, human beings have debated whether man is by nature good or evil. His instincts predestine him to be both. Without instincts he can be neither good nor evil, but only a soulless robot.

In itself, an instinct is neither good nor evil. We may approve of a dog which guards its master's luggage and disapprove of one which bites us in the leg. Nevertheless, we cannot characterize an animal's behavior as good or evil. An intention or an act can be evil only if a person who knows better deliberately harms another person, consciously refusing to exercise what strength he has to inhibit destructive drives.

# PART FOUR

# The Problems
# of Getting Together

# CHAPTER 7

# From Cannibalism to Love

*Overcoming Aggression*

"If we view the matter logically, it is hard to understand why people bother to bear and raise children," writes Walter R. Fuchs. "Bearing children is painful for the woman, and raising them is a lengthy and arduous procedure. A husband must spend years working hard to support his family. Thus if we analyzed sexual relations logically, we humans would never go to the trouble of reproducing ourselves. However, nature robs us of reason so that we behave illogically. Paradoxical as it may sound, if we were not such irrational creatures, our species would become extinct."[1]

Not reason but instinct enables animals to perform their most important life functions. This is true even of the most intelligent animal, man. We inherited our instincts from creatures that lived on earth long before the development of intelligent life. To examine the history of sexual patterns in detail, we must return to the dawn of evolution.

The "invention" of the male created many problems. We have noted that nature had to devise ingenious methods so that males and females could recognize each other. However, once the two partners have recognized and been attracted to each other, their problems have only begun.

Despite their mutual attraction, male and female fear each other, for they are strangers. Human lovers inherited this fear from the lower animals.

In certain species, males and females do not experience a strong mutual fear. In these species, nature had to devise techniques to prevent the mates from murdering each other. In some cases these techniques were ineffective, and mating is attended by cannibalism.

Manfred Grasshoff gives a striking description of a species of spider known as *Araneus pallidus*.[2] After searching for some time, a male *Araneus pallidus* found the web of a female four times his size and began to tug at a signal thread as if he were tugging at a bell-pull. He gave eight short tugs followed by a single violent jerk that shook the female's body. After a pause of several seconds, the male began to "ring" the signal thread again. He went on signalling for five minutes until the female slid down beside the signal thread, exposing her abdomen and twitching her two front pairs of legs as if she were inviting the male to mate with her and at the same time threatening to kill him.

Gradually the male approached the female until he was some 1/50 inch away. The jerking of the thread slowly changed into a trembling of his whole body. The trembling ceased when the male thrust his front pair of legs underneath him and stretched the second pair forward. He paused for a moment without moving. Then he appeared to aim the pedipalp, the leg bearing the sperm capsules, at the female's epigynum, or genital opening. Several times he repeated these movements and then finally lunged at the female. His first attempt to mate with her failed. A fraction of a second later, the male had spun a thread and dropped to a safe distance from the female. Then he climbed back up and began to court her again.

At thirty-second intervals, the male made ten more unsuc-

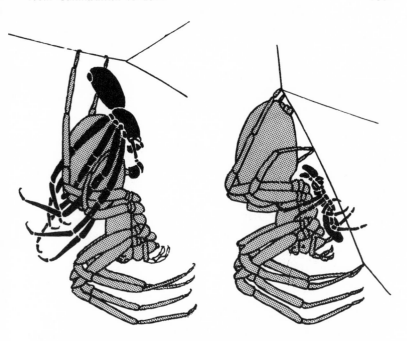

On the right a female *Araneus pallidus* eats the smaller male spider during copulation. The spider on the left is of a different species, (*Araneus diadematus*). Somewhat larger than his unfortunate counterpart, the male of this species generally succeeds in gripping the female tightly so that she cannot eat him either before or after mating. The smaller a male spider is, the greater the danger that he will be devoured on his marriage bed.

cessful attempts before he finally landed under the female, his back touching her abdomen. Then he inserted the pedipalp into the female's epigynum. In this position, his abdomen was directly beneath the female's mandibles. In a moment she dug her jaws into his body. Copulation continued for nine minutes. Then the female jerked her bridegroom forward under her jaws and began to devour him.

Further observation of the spiders and an examination of their physical structure revealed that male and female could not mate unless the female dug her deadly jaws into the male's

## CAPTIONS FOR THE COLOR INSERT THAT FOLLOWS

The female Javan hill mynah bird listens reverently to the male's loving serenade. For the most part his song consists of shrill whistles and hoarse chuckling sounds. Yet despite its limited repertoire as a songster, the Javan hill mynah is one of the most accomplished mimics in the entire bird kingdom. The mynah can pronounce human words more distinctly than a parrot. Monogamous birds that live in large flocks are often very talented mimics. They invent special sounds to call their mates "by name" and summon them from the rest of the flock. A mynah's mate, surrounded by the squawks of all the other birds, nevertheless manages to hear the one sound that is meant for him or her. The ability of mates to communicate with great precision helps to keep a mynah pair together.

Grasshoppers face many problems during courtship. Before the male can mount the female, he must search for her in the seemingly endless jungle of a meadow where he cannot see more than a couple of inches away. Initially, both male and female chirp the song peculiar to their species, so that neither will make the mistake of attempting to mate with a grasshopper of a different species. The male uses the female's song as a guide to her location. Whenever he hears her song, he leaps in her direction. When he is not singing his species song, the male sings a war cry designed to drive away male rivals of the same species. If the male hears a supposed female singing the war song, he knows that it must be a male and gives up the pursuit. If the other grasshopper proves to be a female, both grasshoppers sing a courtship song that gets them into the proper mood to mate.

These moths are known as six-spot burnets. When insect-eating birds see the moths' red wings, they behave like bulls that have seen a red flag and swoop down to capture them. However, the moth's body juices are so poisonous that the bird drops the insect as soon as it takes it in its beak. Any bird that has ever touched a six-spot burnet remembers what the moth looks like and avoids it in the future. Thus the moths feel so secure that they fly very slowly and never attempt to flee or hide. When a burnet is sucking nectar from a blossom, a human being can easily capture it. Moreover, as the picture shows, the moths also dare to copulate in full view of their enemies.

Blue-backed fairy bluebirds, which are native to southeast Asia, are monogamous birds, but they conduct a "long-distance" marriage. The brilliant, colorful plumage of the male might attract the attention of predators to the nest, which is well-camouflaged with moss and lichen. Thus the male never approaches within sixty feet of the nest. Remaining about that far away, he keeps watch, warns his family of approaching danger, and attempts to divert the attention of enemies from the nest. While the female is sitting on the eggs, the male feeds her. However, she has to leave the nest to receive the food, so that the two birds will not betray its location.

The fish known as Chelmon rostratus, a variety of butterfly fish, are monogamous and mate for life. However, once they have spawned, they completely ignore their young. Thus the care of young is not the only thing that can hold two mates together. These fish, which inhabit the coral reefs of the Indian and Pacific oceans, behave very aggressively toward all other members of their species. Once the male and female have overcome this barrier of aggression, it is to their advantage to remain together. If two mates stay together, they do not have to fight to win a new mate each mating season. When erect, the fins on the backs of these fish bristle with quills, which can inflict serious injury.

The picture shows a North American snapping turtle. Land turtles are incapable of communicating their emotions to each other. Thus the male, unable to obtain the sexual consent of the female, is obliged to rape her.

The male American red squirrel courts a female by running her to the ground.

During copulation the male lion stands over the lioness the way he would stand over a slain gazelle and takes the nape of her neck in his jaws. However, he does not actually bite her. Instead he merely nibbles at her a little, as if he were saying, "I love you so much that I could just eat you up!" Lions live in prides and thus are social animals, accustomed to daily contact with other lions. As a result, during mating lions are less aggressive than their solitary and antisocial relatives, the tigers.

body. If her mandibles did not hold him in place, the tiny male would slip off the armor plating of his giant bride.

In a closely related species of European spider, *Araneus diadematus,* the males are only slightly smaller than the females. Lunging toward the female, the male wraps his legs around her like a wrestler and holds her so that she cannot bite him or spin threads to imprison him. Still maintaining his wrestler's hold, he finishes mating in record time—between three and twenty seconds. Then he streaks away and hides on the ground, where he dies several days later. The female *Araneus pallidus* would probably say that allowing a male to live for several days after mating is just a waste of good protein. After all, once he has served his purpose, he is good for nothing but to nourish his bride.

However, for the most part males are not doomed to serve such a utilitarian end. As a rule, predatory insects can find enough to eat without devouring their mates. Among the thirty thousand species of spiders, cannibalistic mating is the exception, not the rule. Thus clearly nature did not design everything to serve a useful purpose. In fact, it is in no way "normal" that a male should be attracted to a creature which will inevitably kill and devour him. In this respect, the behavior of the male *Araneus pallidus* is quite exceptional.

Cannibalistic mating did not develop because it helped to guarantee the preservation of the species. Rather it was an anomaly which persisted because it did not actually prove disadvantageous. In the majority of spider species, the females do not devour their mates. This proves that reproduction is more efficient when it does not require self-destruction. The male *Araneus pallidus* is justly terrified of his murderous bride. Thus if the species is to survive, the male's desire to mate must prove stronger than his fear of death. As a rule, the instinct of self-preservation is stronger than all other instincts. Thus in the history of evolution, it clearly proved more efficient to devise techniques to protect the male during mating than to devise males capable of overcoming the fear of death.

Besides the wrestling hold of *Araneus diadematus,* nature has devised many tricks to protect males during mating. The jaws of the male trapdoor spider are so constructed that he can lock them between the jaws of the female to keep her from biting.

A male spider belonging to the *Pisauridae* cleverly outwits his mate by capturing an insect, wrapping it up in silk threads, and bringing it to the female as a gift. While she is busy eating her gift, he hastily mates with her and departs. One male of this spider group is so nimble that he finishes mating before the female has begun to eat her insect. Then he hurriedly grabs the gift and absconds with it. Thus the female has neither her gift nor her husband to eat at her wedding-feast.

The most famous cannibalistic females belong to the order of mantises. Many are members of the praying mantis family. Robert Burton described one bizarre species of praying mantis. An inhibitory mechanism lodged in the nervous system of the head (these creatures do not really possess brains) prevented the male's sexual organs from expelling sperm. No other inhibitory reflex in the entire animal kingdom functions as flawlessly as this. Only when the neural mechanism is destroyed, i.e., when the head has been cut off by a razor blade or bitten off by the female, can the male mantis release his sperm.

When the male mantis glimpses a female, he "freezes." Then he begins to creep up on the female from behind, keeping to the left side of her body. He moves so slowly that a human being can detect his movements only by filming them with a quick-motion camera. It takes the male mantis an hour to move an eighth of an inch. Any sudden movement would mean instant death. The female praying mantis is an insatiable cannibal and will devour the male the moment she sees him, without giving him a chance to mate with her.

When the male is close to the female, he springs at her. If he does not judge the distance correctly, he will be devoured at once. If he lands in the right spot, he will be eaten in two stages. As soon as the male is in position to copulate, the female bites off his head, removing the neural mechanism which prevents him from releasing his sperm. Reflexes cause the headless body to fertilize the female. Once the sexual act is ended, the female eats the rest of the male, leaving only his wings.

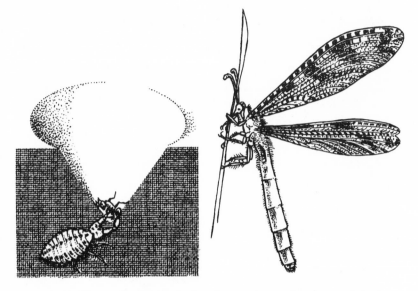

At the left an ant lion in the larval stage has captured an ant in a small conical pit. Eventually the larva develops into the beautiful winged creature on the right, which is also a voracious cannibal.

Like female praying mantises, female ant lions do not believe in equal rights for males. When the larvae mature, they are no longer "big-bellied dwellers in dark traps" (Richard Gerlach[3]), but beautiful winged creatures. As adults they look so ethereal that until recently, scientists believed that they must live on honeydew.

In reality, adult ant lions, like their larvae, are insatiable predators. The females are even more predatory than the males. Not only do they eat most species of flies and insects, but they eat their own bridegrooms immediately after mating. Then they eat all the other males which make the mistake of approaching them. They do not even bother to mate with these latecomers before eating them; for having mated once, it is unnecessary for them to do so again.

Two factors may lead female praying mantises and female ant lions to devour males. For one thing, they have a voracious appetite. When praying mantises are enclosed in a terrarium and given an abundant supply of live insects, they never stop eating. Secondly, mantises and ant lions regard all other insects as potential victims. They will eat anything that does not have the strength to resist them. The IRM which enables the female to recognize prey does not allow her to distinguish between her mate and other insects; thus she treats him as she does any other prey.

The females of some species of biting midges are equally the slaves of their instincts and thus cannot distinguish between males of their species and other prey. Biting midges are annoying little flies less than one millimeter in length, which can slip through mosquito netting and inflict painful bites. There are many species of biting midges. Instead of biting human beings and other mammals, the females of some species eat various types of flies, including males of their own species.

Swarms of male flies, gnats, and midges gather in ditches, along forest paths, beneath trees, near bushes and stones, and even near moving objects like cows or people. The little clouds of dancing males attract the attention of females which are willing to mate, and the females fly over to join them.

Clouds of male midges gather together to dance like male flies of other species. But female midges belonging to species that hunt other gnats and flies are indifferent as to whether a cloud of males is composed of members of their own species. These females are not interested in mating but only in prey.

The cloud of males becomes a scene of carnage. When a female reaches the cloud, she soars down onto a male's back, grips him tightly with her legs under his belly, digs her jaws into his head, and injects a strong corrosive saliva beneath his armor plating. In a short time his body dissolves, and the female sucks it out of his armor the way we suck milk from a bottle through a straw. Afterwards she throws away the empty husk.

The female midge eats all male gnats and midges, including males of her own species. If she devours a male of her own species, he posthumously becomes her mate. When she throws away his empty armor, the end of his body, which contains his sexual organs, clings to her abdomen and fertilizes the female.

Meanwhile the male himself, reduced to liquid form, is flowing through her body.

Of course, neither the male nor the female midge understands what is happening. Only human beings are capable of associating the sexual act with the production of young. The instincts of the male midge compel it to dance in a swarm with other males. It never actually hunts for a female. The female in turn is not thinking about sex, but only about hunting. She has no conception of what a male of her species should look like. She only knows how to recognize prey. The image of her bridegroom acts as a key stimulus, triggering her impulse to hunt. She has no mating drive. The act of mating is a mere reflex, a by-product of the drama of eating and being eaten.

The mating habits of some mammals border on cannibalism. For example, rhinoceroses, which are armored like the insects, conduct an extremely violent courtship. Until quite recently, no zookeeper dared to leave a male and female black rhinoceros together during mating season.

In 1968 in a zoo in Chester, England, thirty zoo attendants surrounded a pen housing a female rhinoceros named Susie.[4] Anticipating that the male might try to kill the female, the men armed themselves with firehoses, long poles, and chains. Then the sliding door to the neighboring pen was raised. Susie gave a shrill whistle. Suddenly four thousand nearsighted pounds of muscle, covered with thick horny skin and armed with two lowered horns, came streaking toward Susie's pen at a speed of almost twenty miles per hour. It was Roger, the bull rhinoceros. At the last minute, Susie turned aside, allowing the male to charge past her.

This was the pair's first declaration of love. Then Susie charged at Roger. She managed to get her horns under his four-thousand-pound body, tossed him two feet in the air, and let him fall with a crash. Seconds later, agile as a weasel, the bull rhinoceros was back on his four stumpy legs. At once he rammed Susie so hard that the blow would have badly dented a car.

Uttering blood-curdling whistles, squeals, and grunts, the two animals went on ramming each other for another hour. At any moment one of them might have succeeded in trampling the other to death. Several times the keepers prepared to turn on the firehoses to separate the two giants. Thus instead of a tender courtship leading to intimate union, the male and female black rhinoceros engage in a brutal conflict which endangers both their lives.

In the wild, rhinoceroses are solitary creatures hostile toward other members of their species. Normally, they simply avoid each other and thus do not fight. However, during mating season sexual attraction forces them together, releasing their aggressive impulses. At first male and female appear to be deadly enemies.

Finally the keepers observing Susie and Roger noticed that the animals were no longer attacking each other as ferociously as before. Then Susie began to trot around the pen as coquettishly as her clumsy body permitted. She concluded her performance by inviting the male to mount her.

At once Roger climbed on top of her, maneuvering until all four of his legs were off the ground and Susie was staggering under his weight. She managed to hold him up for an hour, until he finally slipped off her back again. At that point, both animals were too tired to fight.

Two factors have prevented rhinoceroses from developing a more refined mode of courtship. For one thing, their clumsy bodies make it impossible for them to communicate with subtle gestures or facial expressions. Secondly, in most cases their horny armor prevents them from being hurt. When both partners are well-armored against blows, there is no need for them to treat each other with gentleness. Nevertheless, observation of rhinoceroses on the East African Serengeti Plain has revealed that sometimes their violent courtship results in the death of male or female.

Land tortoises also conduct a brutal courtship. However,

An affectionate pair of jackdaw mates. The male, on the left, has irritated his mate by stealing food from her. Now he is appeasing her wrath by assuming the pathetic posture of a baby jackdaw.

their courtship ritual is much more leisurely than that of the rhinoceros. Like the male rhinoceros, the male tortoise has to find some way of getting the female into the proper position for mating. He has devised a primitive but effective solution to the problem.

During mating season, the female tortoise becomes so fat that when she is in danger, she cannot get all of her body underneath her shell at the same time. Either her head or her rear parts must remain outside the shell. The male tortoise attacks her until she is forced to withdraw her head, leaving her genital opening exposed at the rear.

The male tortoise begins by ramming the female in the side or from behind, then pursues her, biting her forelegs. Both male and female move at what for tortoises is breakneck speed. Frequently the "game" continues for hours. Sometimes the female loses so much blood from her many wounds that she dies.

Eventually the terrified female gives up the attempt to defend herself and remains motionless with her head and forelegs inside her shell. Then the male can safely approach her rear. In effect, the male rapes the female. Unlike rhinoceroses, the two tortoises do not exhibit even a flicker of affection for each other.

Male frogs also rape the females. Frogs were the first vertebrate land animals to sing love songs. At night the males perform croaking concerts designed to attract females of their species. But apart from their songs, the males make no effort to be charming. As soon as the female swims close enough, the male halts his courtship and seizes her in a wrestler's grip. A male frog will jump at anything that is approximately the size of a frog, grip it with his forelegs, and squeeze until the spawn is expelled from the female's body. Then he fertilizes the spawn.

Squacco herons, which are among the most beautiful of all long-legged birds, are found in southern Europe, Africa, and Asia Minor. Males and females have equally beautiful plumage. Squacco herons are monogamous and mate for life. During mating season their yellow beaks turn dark blue and their yellow-green legs become red.

He may rape a female of his own or another species, or even a toad.

Groups of male frogs engage in mass orgies of rape. Yielding to mass hysteria, a male will jump onto the back of another male, or even a stone or a piece of wood. Two males may jump on the same female, or a male may leap onto the back of a male which is riding a female and ride him piggy-back.

Male frogs which are seized by other males struggle violently and eventually manage to get free. Moreover, a male will usually drop a stone or a piece of wood once he realizes that it does not look or feel like a frog. However, a lump of mud may feel very much like a female, modelling itself in the grip of the male until it takes on the shape of a frog. Thus a male may hold onto the mud for days, vainly waiting for it to release the spawn. He will also hold onto a dead frog or a female too young to produce spawn.

The males of one species of African clawed toad are extremely powerful. Although only half the size of the female, the male toad can roll her around on the ground and squeeze

The male of a species of African clawed toad is only half the size of his mate; yet he manages to squeeze her so hard that she loses consciousness.

her so hard that she loses consciousness. After the female is unconscious, the male squeezes out the eggs and fertilizes them.

For over three hundred million years, rape has been the frogs' normal method of mating. The reason for this may be that frogs do not actually copulate. That is, the male does not insert a sexual organ into the female. Instead he clings to the female's back until she releases the spawn, which he fertilizes directly. This being the case, it seems natural that the male and female should synchronize their movements so that the male helps to press the eggs from the female's body and is there to fertilize them the moment they emerge. The act of aiding the female to release her eggs could easily lead to a form of rape.

To engage in genuine copulation, sexual partners must be on more intimate terms than male and female frogs. However, there are exceptions to this rule. Exceptions occur when the male is physically much stronger than the female, as is the case with the Northern elephant seal.

A female Northern elephant seal weighs almost two thousand pounds and thus is certainly no pygmy. However, the male is four times her size and weighs almost eight thousand pounds. The cows appear to be afraid of these colossal lumps of flesh and fat. Whenever she gets a chance, a cow leaves the harem she shares with some ten to forty other females and wanders along the shore of her Northern Pacific island home, looking for a bachelor lighter in weight than her husband. Usually the bull attacks the fleeing female, biting her and beating her so savagely with his flippers that in future she is too afraid to run away again.

When the bull is in the mood for sex, he grabs one of the females in his harem, wraps one flipper around her in a wrestler's grip, and smothers her beneath his weight, curving his rear end around so that the two can copulate. After five minutes, he lets the female go. Often she succeeds in escaping even sooner, screeching, kicking, and beating at the sand to free herself from the weight that is crushing her.

The males of some smaller mammalian species, such as male squirrels, behave with equal crudity. The female does nothing to invite attack except to utter a mating call. The sound carries almost a thousand feet into the surrounding forest. All the males possessing territories in this area, as well as young males which have not yet staked out a territory, respond to the call and begin to hunt down the female. As many as twelve males may set out in pursuit of a single female.

At first the female squirrel tries to drive the males away; but being badly outnumbered, she soon has no recourse but flight. The males chase her up and down trees through underbrush and clearings. The hunt may continue for hours or even days. Hunted and hunters alike travel farther and farther away from their home territory.

Whenever the female encounters one of the males, the two of them scuffle and threaten each other. The males also fight among themselves. The strongest male bites his rivals until they begin to hang a discreet distance behind him.

The female squirrel will not mate until she is totally exhausted. At this point, the male begins to demonstrate a little tenderness. He is forced to court the female, for he cannot simply rape her. The female's bushy tail covers the genital opening and cannot be removed unless she chooses to remove it. The tail acts as a sort of built-in chastity belt. Thus once the female is so exhausted that she cannot run or fight back, the male has to charm her into lifting her tail.

The male courts the female by squeaking plaintively like a baby squirrel, thus arousing the female's maternal instinct. Eventually she responds to the childlike, seemingly helpless male. Thus the sympathy bond between squirrel mates is rooted in the adult female's bond with her young.

Many adult animals besides squirrels exhibit childlike behavior during courtship and mating rituals. Moreover, human beings too are biologically predisposed to behave like children when they are in love.

When the female squirrel is ready to mate, she signals the male by releasing several drops of urine and raising her tail. Then the male gets into position, raising his tail in the shape of a question mark over his back.

After mating, the male and female squirrel remain together for some time. For several days they even inhabit the same nest. But long before the young are born, the mates lose their tender feelings for each other. Finally the female drives the male away.

Nevertheless, for a brief period the two squirrels feel something for each other that persists beyond the sexual act. They have taken a first step in the process of socialization that leads to marriage.

Only when there is a strong sympathy bond between the sexual partners will their relationship end in genuine marriage. However, to varying degrees, many animal mates develop sympathy bonds with their mates. In many cases their feeling for each other dies soon after mating. Sometimes it dies so abruptly that the erstwhile mates are in danger of killing each other. For example, this is the case with the tiger.

Unlike a rhinoceros courtship, a tiger courtship begins with tenderness and ends in a murderous frenzy. Tigers are solitary animals which are hostile toward members of their own species. Thus like the male squirrel, the tigress breaks down her partner's resistance by behaving like a baby—in this case a tiger cub. As soon as she sees a male, she rolls over on her back, kicks her legs in the air, and miaows plaintively. The male approaches and observes her from a safe distance. Then the tigress slinks around him, playfully arches her back, purrs, and finally rubs the male's face with her chin whiskers.

However, the male tiger is mistaken if he thinks that the honeymoon has begun. As soon as he tries to respond to her caresses, the tigress abruptly changes from a kittenish, purring cub to a snarling beast with gnashing teeth and lashing claws. The tigress is not simply employing feminine wiles, pretending to reject the male in order to attract him all the more. Instead, her ferocious response indicates that she has not yet been able to completely overcome her aggressive instinct. In view of the highly aggressive nature of tigers, this is hardly surprising. The moment the male tiger begins to respond to her advances, the tigress seems to forget that she initiated the courtship. Suddenly she sees the male as an enemy. The two animals continue to flirt and negotiate until they both become sexually aroused. Sexual arousal, coupled with the bonding instinct, enables them to overcome their aggressive drive and permits them to touch each other, paving the way for the sexual act.

During the actual act of mating, the male tiger appears to

behave aggressively toward the female. Mounting the tigress, he takes the nape of her neck between his teeth as if he were about to kill her. However, in reality he will do no more than nip her slightly.

The sexual relations of almost all animal species exhibit elements of aggressive behavior. The courtship of the tiger represents a fusion of aggression and attraction. The feeling of attraction acts as a brake to prevent the two animals from killing each other. Love and aggression are never very far removed. For example, human beings often tell each other, "I love you so much that I could just eat you up!"

After the two tigers have mated, for the first time the male tiger is in real danger. The sexual desire of the tigress abruptly vanishes, and she turns in fury on the alien creature which has clambered onto her back. The feeling of attraction which braked her aggressive drive is now gone. Shaking the male off her back like a few drops of water, she springs at him with murder in her heart.

The male has no recourse but to flee. He is stronger than the female and could easily defend himself against her attack, were it not for the fact that his aggressive drive is still blocked. The female's attraction to the male disappears immediately after mating; but the male's attraction to the female persists, rendering him helpless. He is emotionally incapable of defending himself. Thus unless he runs away, he will be killed.

In most older zoos, the animals live in cages as small as prison cells. In these cages, a tigress can kill her lover within seconds after they have mated. Thus in older zoos, tigers were not allowed to mate and never produced offspring. However, in modern zoos the animals live in large enclosures. Here the male tiger can run away after mating and thus save his life. In the Hagenbeck Zoo in Hamburg, Germany, the tigress cools her wrath by taking a dip in the pond after her mate has fled.

Immediately after attacking the male, the tigress in the Hamburg zoo begins to miaow plaintively and behave like a young cub again, trying to attract the male's attention. For the brief time she is in heat, the tigress may invite the tiger to mate with her no less than eighteen times a day. Yet after each mating, she once again tries to kill him.

Tigers are extremely aggressive and solitary animals. They do not band together in groups and do not ever "marry," i.e., live permanently with a mate. They know nothing of the personal relationships which play so large a role in human life, enabling two people to retain their affection for each other after the completion of the sexual act. Tigers can experience only two feelings for each other: sexual attraction and the desire to kill.

Lions copulate much as tigers do. That is, the male takes the female's nape between his teeth. However, a lioness will never attack a male lion. Unlike the solitary tiger, the lion is a social animal. The bonding instinct prevents a lion from killing other members of its pride.

The golden hamster resembles the tiger in that its mating habits represent a fusion of attraction and aggression. Human beings are amused by the antics of these little creatures. However, they are by nature quarrelsome and solitary. Males and females do not associate except during mating season. Several days after they have mated, the female attacks the male and drives him away.

My family owned a pair of golden hamsters named Fridulin and Minchen. Nine days after they mated, they still appeared to be the most loving couple in the world. Sitting next to the food bowl, Fridulin munched a mouthful of food, meanwhile gazing with large, faithful eyes at his mate. Suddenly Minchen fluffed out her golden fur, trotted over, bared her two front teeth, and snarled savagely in Fridulin's face. Terrified, the male darted into the corner and remained there without stirring. When a caged female hamster becomes aggressive, the larger, more muscular male must be removed to a place of safety; for the male is emotionally unable to defend himself and can be torn to shreds. In their native habitat, the Syrian desert, the males are free to run away, just as the male tiger can escape the tigress by fleeing into the Indian jungle.

In many species, mating is followed by the resurgence of aggression. However, in species like the golden hamster, the mates do establish a bond which temporarily persists beyond the sexual act. Such species represent transitional forms between species in which the partners develop no personal bond and those in which they engage in more permanent kinds of pair formation.

In the course of evolution, nature has developed various methods of curbing the aggression of two animals so that they can tolerate the bodily contact involved in the act of mating. One obvious solution was to devise techniques by which two animals could mate without ever touching each other. The males of many arthropod and amphibian species enclose their sperm in one or more packets called spermatophores, which they leave around for the females to find. The females must hunt for the packets the way children hunt for Easter eggs, and insert the sperm into their bodies. Thus long before man, nature began the practice of artificial insemination.

However, this approach is less effective than it may appear, for many sperm packets are never found by the females. To keep the loss of packets to a minimum, it proved advantageous for male and female to come into close contact. The family of newts and salamanders shows the variety of methods employed by species that reproduce by means of spermatophores.

The newt called *Ranodon sibiricus* developed a very primitive mating technique. The males and females never come into personal contact. The male attaches his sperm packet to the bottoms of stones lying in the water. Here his responsibility ends. Eventually a female comes along, sniffing at the stones. When she finds a spermatophore, she attaches to the same stone a small sac containing twenty-five to fifty eggs. Soon the sheaths covering the two packets dissolve. The sperm fertilizes the eggs and a new generation begins to grow.

Thus the female *Ranodon sibiricus* never feels attracted to the male of her species. If she can be said to love anything, it is only the packet of sperm.

Males and females of the genus of true newts have a more personal relationship. The male deliberately crosses the female's path, then moves along in front of her, depositing packets of sperm in her path. Depending on her capacity for putting notions into action, and on what mood she is in, she may pick up the packets and insert them into her cloaca, walk right over them, or simply devour them.

Among the Asian land salamanders, the female takes the initiative. Finding a puddle of melted snow high in the mountains, she waits until several males are near by to observe her and then attaches her egg sac to a stone in the puddle. While she is still busy attaching the eggs, a male will come over, push her away with his hind legs, and wrap the sac in the threads containing his sperm. Despite the momentary personal contact between the mates, theirs is still a highly asocial method of fertilizing eggs.

Mole salamanders have their own technique of reducing the loss of sperm packets. Unlike the male newt, the male mole salamander does not run along in a straight line depositing packets in front of the female. Instead, both male and female crawl around in a circle dancing the "salamander waltz." As they dance, the male releases his packets. The female may overlook many of the packets, but at least she has a better chance than the female newt of finding at least one.

Mole salamanders developed the first courtship dance in the evolution of life on land.

The male Alpine salamander deposits his sperm packet near a female and then leads her over to it. The male fire salamander is even more cautious. First he rides around on the female's back, then slips underneath her and carries her piggyback. Finally he places the sperm packet directly beneath her cloaca, so that all the female has to do is insert it into her body. Thus although fire salamanders do not actually copulate, they will tolerate each other's touch.

Male and female two-toed Congo eels (amphiumas) and European mountain salamanders not only tolerate each other's touch, but actually copulate. After a ritual courtship, the male inserts the sperm packet directly into the female's genital opening. Since these species copulate like birds and mammals, the males do not really need to enclose the sperm in packets. Moreover, the female appears to feel attracted to the male which produces the sperm as well as to the sperm itself.

To prevent the loss of sperm packets, nature ultimately devised copulation, a new technique for fertilizing eggs. In the history of evolution, many paths have led to this same goal: a form of mating involving physical contact and physical union.

Like all other animals, amphibians face the problem of how to overcome their dread and dislike of physical contact.

Water-dwellers practice methods of fertilization that spare them the necessity of physical contact. Many species of bivalves play a "game of chance." Masses of oysters live anchored to oyster beds, periodically releasing their eggs and sperm into the sea. Chance dictates whether the sperm will fertilize the eggs.

Oysters can improve their chances in this lottery by producing vast quantities of "lottery tickets," i.e., sperm and eggs. A single American oyster can release as many as one hundred million eggs at a time. Moreover, the millions of oysters in an oyster bed can arrange to release their eggs and sperm simultaneously. Clouds composed of trillions of eggs and sperm float around beneath the sea. Under these conditions, the eggs have a good chance of being fertilized.

Apart from these orgies of expelling eggs and sperm, oysters have no sex life. The males and females cannot get close to each other, for each oyster is rooted to one spot. The byssus

During courtship, two bee-eater birds do not know which of them is the male and which the female. Sometimes two courting birds may be of the same sex. They take turns playing the male and female roles until each bird finds out which role is more comfortable. If the two birds are compatible, they must be of different sexes. Then they become mates. However, they do not copulate until they have dug their nest hole together. Although the birds in the picture have been mates for some time, the female (on the left) will not allow the male to mount her and mate with her until he has made her the gift of a bee.

Although hares may appear to be amiable, good-tempered animals, they engage in a rather violent courtship. Except during mating season, they are solitary and quarrelsome creatures. During mating season they are extremely promiscuous. When the female is in heat, she flees from the male, at the same time twitching her black-and-white tail, which sexually arouses the male. When he has cornered her, he beats her until she submits. Male and female separate immediately after mating. Their sexual desire is quickly extinguished, and no personal bond exists to keep the mates together. Thus hares are incapable of marriage.

gland of oysters produces a gluelike substance with which they anchor their lower valves to the rocky ocean floor.

However, although oysters cannot move around, they do have the ability to repeatedly change sex. When they reach sexual maturity, all oysters are males. After releasing their sperm, it takes them several weeks to change into females. Then they release their eggs and in a few days change back into males.

A mysterious physiological clock makes all the oysters in an oyster bed change sex at the same time. This clock is governed by the moon. In the period between June 26 and July 10, two days after the full or new moon—i.e., during the spring tide —the sperm of the males and the eggs of the females are fully developed. All the oysters wait for the signal to release them.

Zoologists conducted experiments to see how this signal was issued.[5] They discovered that every egg laid by an American oyster emits a scent which acts as a signalling device. As soon as the perfumed water reaches a (temporarily) male oyster, he releases his sperm, which is also scented. The scent of the sperm in turn signals females in the area to expel their eggs. Thus as soon as a single egg is expelled, its sets up a chain reaction. All the oysters in the bed follow suit and release their eggs and sperm in rapid succession.

The simultaneous release of vast quantities of eggs and sperm helps to ensure the survival of the American oyster. It's cousin the European oyster enhances its chances of survival not by "mass production," but by caring for the eggs until they are fertilized. The female European oyster does not lay one hundred million eggs, but "only" around one million. She incubates the eggs in a special area of the mantle cavity inside her shell. Every hour, between two and four gallons of water

These three grass snakes are caught in a love triangle. They have tied themselves in a knot, forming a "Medusa's head." During copulation a male and female snake may remain joined together for hours or days. The female can retain the male's sperm and keep it active for as long as five years. Each year she uses part of the supply to fertilize herself.

flow through her gills. Her food consists of protozoans and organic matter that she filters from this water. Frequently the water also contains male sperm cells. The oyster's sensitive sensory apparatus detects the sperm, which is then conveyed not to the stomach but to the eggs. Oysters cannot move from their oyster bed. Thus instead of trying to attract the male to her side, the female simply collects his sperm and uses it to fertilize her eggs.

Many fish reproduce by the same technique as the American oyster. Fish are not anchored to one spot like oysters in an oyster bed. However, at spawning time masses of male and female fish assemble at traditional spawning grounds, where almost simultaneously they expel their eggs and sperm. Thus no physical contact occurs between the fish, but only between the eggs and sperm.

As is the case with oyster eggs, chance decrees whether or not the fish eggs will be fertilized. To improve their chances of survival, the fish must mass-produce their eggs. The ling fish belongs to the cod family. At spawning time, the female ling lays between sixty and eighty million eggs. Once scientists observed a female which laid one hundred and sixty million eggs.

Assuming that the ling population in a given region remains relatively stable, this means that of the eighty million possible offspring of a female ling, only two survive until maturity. Either the eggs are never fertilized, or the embryos die or are eaten, as are most young fishes. Nature is a harsh mistress.

This random method of reproduction is not very efficient. However, in other fish species, males and females come into closer contact.

European grayling fish spawn in shallow, swift-flowing mountain streams. They are very suspicious fish, and if they see a shadow on the water, they will immediately flee for cover. To observe them, one must lie beside the bank of a stream, camouflaged with reeds, and carefully peer out at the fish with binoculars.

Accompanied by a film team, I observed several female graylings use their fins to dig holes four inches deep in the pebbly bed of the stream. The holes were dug to receive the spawn. Then the females waited for the males to arrive. However, no males appeared, for we had captured and removed them all.[6]

Earlier we had placed an oar in the water near the female graylings and their holes. Now we cautiously shook the handle so that the blade of the oar began to vibrate. The fish did not flee from the oar. Instead, the female which had dug her hole just beneath the oar swam closer, turned to the side so that her body was parallel to the blade, and began to deposit her three to six thousand eggs. Clearly she had mistaken the vibrating oar for a male grayling.

Normally the males appear when the females have prepared the spawning holes. At first they fight among themselves to drive away rivals. When all his rivals are a safe distance away, a male will swim to a female, line up his body parallel to hers, and begin to tremble. The female, only an inch or so away, begins to tremble too. Without ever touching each other, the two fish simultaneously release spawn and sperm into the water. The swirling water mingles the spawn and sperm, fertilizing many of the eggs, which sink into the hole beneath the fish.

The experiment we conducted reveals what an "abstract" conception a female grayling has of the male. She interprets anything which vibrates near her to be a male, and thus cannot tell a male of her species from a wooden oar. When mating does not involve bodily contact, the physical shape of the male is of little importance.

The closer male and female come during fertilization, the more eggs will be fertilized by sperm cells. Thus it is advantageous for mates to engage in intimate personal contact.

In spring, the haddock travels to its spawning grounds off the coast of Norway and in the northernmost regions of the North Sea. Some three hundred and twenty-five feet beneath the sea, the males beat their "courtship drums." That is, they make their air-bladders resound. The drumming attracts females and drives away male rivals. The males engage in drum-beating duels, and the male that can drum loudest and longest is the victor.

When a female approaches the victorious male, he accelerates the tempo of his drum-beats to create a humming sound.

At the same time, he ostentatiously beats his fins, performs acrobatics, and repeatedly changes the color of his scales. If the female likes his performance, she begins to follow him.

Pausing in his courtship dance, the male taps the female's side with his head. This is the signal to mate. The two fish spin around and lie belly to belly. In this position they rise straight up for a distance of from thirty to sixty feet, at the same time releasing their eggs and sperm. What a contrast to the "abstract love" of the European grayling!

When bodily contact occurs during mating, the "person" of the male becomes important to the female. The male engages in display behavior, in a courtship ceremonial to please his potential mate. The ceremonial is designed to overcome the natural antipathy existing between two members of a species and to enable them to establish a personal bond.

# CHAPTER 8

# Am I Male or Female?

*Confusion about Sexual Roles*

A science-fiction novel tells of a group of astronauts whose spaceship lands on a distant planet in the year 2074. On the alien planet they encounter strange creatures which reproduce every four weeks. Each creature bears one hundred babies with which it nourishes its cannibalistic appetite. In this distant world males are almost superfluous, for each sexual act produces eight hundred young. Moreover, the longer these remarkable creatures spend fasting, the longer they live, and every attempt at cleanliness and hygiene swiftly results in death.

We do not have to wait until 2074 or make a journey to distant planets in order to meet creatures as bizarre as these. We can observe them in an aquarium in the comfort of our living rooms. Everyone has seen the fish called guppies. These amazing fish possess the same traits as the mysterious creatures from the science-fiction novel.

C. M. Breder, curator of fishes at the American Museum

of Natural History and former director of the New York Aquarium, conducted an experiment to determine why guppies produce and devour so many young.[1] Breder equipped an aquarium with food and oxygen to support five hundred fish. Then he placed a pregnant female in the tank. Over the next six months, she gave birth on four separate occasions, producing batches of one hundred and two, eighty-seven, ninety-four, and eighty-nine young, or a total of three hundred and seventy-two. However, at the end of six months, only nine of the young—six females and three males—were still alive. The mother had devoured all the others shortly after birth.

In a second aquarium of equal size, Breder placed seventeen males, seventeen females, and seventeen baby guppies, a total of fifty-one fish. Soon the females gave birth to large numbers of offspring. Immediately the "Establishment" gobbled up every one. Moreover, the adults ate the seventeen young fish which had been placed in the tank at the beginning. Then a mysterious ailment began to strike down the adults. There were food and oxygen in the tank for five hundred fish. Yet at the end of six months, only six female and three male guppies were still alive.

Thus guppies practice a rather barbarous form of population control. They always allow enough of their young to survive to maintain the guppy population at a constant level. Their cannibalism results not from hunger, but from the feeling of being crowded by other guppies. Human beings living under crowded conditions feel pressured, competitive and anxious. Afraid that they may not get what they want, they are driven to desperate measures to eliminate the competition. Crowded guppies must feel much the same way. If a full-grown guppy does not have at least half a gallon of water to swim in, it becomes a cannibal.

Guppies eat their young selectively, leaving two females for every male. Moreover, two female guppies are born for every male; and in every guppy society, the proportion of females to males is two to one. We have no idea how the fish know when to eat a male and when a female baby.

Female guppies are dominant in other respects besides

number. A male guppy grows to be little more than an inch in length. The females are twice this length and weigh eight times as much. On the other hand, the females are a rather drab yellow-gray in color, whereas the males are quite handsome. In their courtship dance, male guppies wave their colorful fins like flags or point them like swords.

There are few species in which the female is the more beautiful sex.

All year long, male guppies display their beauty to attract the females. Their effort to be physically attractive suggests that guppies mate by direct physical contact. This is in fact the case. Unlike most fish, guppies engage in genuine copulation.

Guppies belong to the live-bearer family. In this family, the eggs must be fertilized inside the female's body. However, the male guppy does not possess a penis. Instead, the anterior "spines" of the anal fins on his underbelly form a groove along which sperm travels into the female's genital opening.

When male guppies first reach sexual maturity, they do not know what female guppies look like. At first they display to every other fish that approaches them. We have noted that the females of certain species of biting midge possess no IRM to signal the presence of a mate; thus they perceive all gnats and midges as prey. The male guppy is just the opposite: He perceives every fish of a certain size as a desirable female.

Guppies inhabit the ponds and inland lakes of Venezuela and the islands of Trinidad and Barbados. In the wild, male guppies must often pay with their lives if, instead of a female guppy, they mistakenly court a predator of another species. If they are lucky, they woo a plant-eater, and discover their mistake as soon as they try to mate with it. Thus male guppies have to learn the hard way how to recognize a female of their species. At times the "stupidity" of the males has disastrous consequences and all the males in a given region are devoured. Meanwhile, the females live on, well camouflaged by their drab coloration.

When there are no males, female guppies can manage to get along without them. The female's body contains a "storeroom" for male sperm. The sperm she collects at one mating will last for nine months. During this period, she can spawn eight times and produce a maximum of eight hundred young.

Moreover, female guppies can survive indefinitely without males. After several months, the mateless females begin to develop male sexual organs. Still retaining their ovaries, they become hermaphrodites, fertilize themselves, and give birth to female young. On an average guppies live for only two to three years; but occasionally they reach the age of seven. Very old females sometimes change into males and develop colorful, graceful fins.

Female guppies developed the ability to store sperm and to change sex in consequence of the males' inability to know instinctively what a female guppy looks like. Complex processes involving the central nervous system enable an animal to recognize a sexual partner instinctively. Many species do not possess a sufficiently complex nervous system to recognize the total image of their partner. Like the guppy, they may altogether lack a "partner schema"; they may respond to a simple, abstract symbol or configuration of symbols; they may learn the image of the partner through imprinting; or nature may devise some special trick to enable males of a species to recognize the females.

Zebra finches are among those species in which the males require special training to recognize the females. Human beings can easily tell the difference between male and female zebra finches. The female is quite plain. Her breast is light gray, her back light brown. She has a bright red beak and a pattern of three stripes—black, white, and black—beneath each eye. The male's body is also light gray and light brown. However, in addition to stripes under his eyes, he has a rust-colored spot on each cheek. A black-and-white ripple pattern covers his throat and the upper part of his breast, and a broad rust-colored stripe dotted with white polka dots extends along the side of his body beneath the wings.

To a human being, one animal looks much like any other of the same species. For example, we find it difficult to tell one goose from another. However, most animals can tell the difference between individuals of their own species. Young zebra

The male zebra finch on the left looks quite different from his mate. Nevertheless, young zebra finches cannot tell the difference between males and females.

finches are an exception to this rule. They cannot even tell the difference between a male and a female.

If young male zebra finches are removed from their parents before they are thirty-five days old, they will never learn to tell the difference between males and females. When the males are between thirty-five and thirty-eight days old, the father bird, which has hitherto been a conscientious parent, begins to view his sons as rivals for the favor of his mate. Then he pecks at them and drives them out of the nest.

From this moment on, the young males feel a violent antagonism toward any bird resembling their father. That is, they expand their Oedipus complex to include all male zebra finches. Their fathers teach them to distinguish between males and females, thus sparing them much confusion and inconvenience later on.

Bee-eaters have to deal with problems which zebra finches are spared. Not only do the adult birds not know what their sexual partners should look like, but they do not even know whether they themselves are male or female.

In the spring, flocks containing hundreds or even thousands of bee-eaters leave Africa, where they have spent the winter. At the end of April or the beginning of May, the birds arrive in their breeding grounds in Spain, southern France, or on rare occasions even in Germany. Despite their resemblance to swifts, these coraciiforms are most closely related to the kingfisher family.

As soon as the birds arrive from Africa in the spring, they begin to look for mates. Sitting on the branch of a tree, a bee-eater will wait until another bee-eater joins it. If more than one other bird alights on the branch, the first bird will drive the others away until it is left alone with one partner. Bee-eaters do something rare in the animal kingdom: they defend a small territory which they use exclusively for courtship, never for copulation, eating, nest-building, or raising young. For a bee-eater, courtship is such a complex and difficult process that the bird must have a quiet place to conduct its investigations into its own and the other bird's sex.

Male and female bee-eaters look exactly alike. Since neither bird knows which is which, the two birds must take turns playing both roles. If one bird plays the male, the other must automatically play the part of a female, and vice versa.

The two birds sit close together on the branch and gaze at each other. Then they spread their head feathers and jerk their heads in the air. Each time they jerk their heads, the pupils of their eyes contract and the iris glows a bright red. In effect, they are throwing fiery glances at each other. Occasionally each bird lunges with its beak and kills an imaginary bee on the other bird's breast.

A bee-eater which is playing the role of a female must sometimes stoop down and assume the posture of a female inviting the male to mate with her. This gesture is merely a symbolic part of the ritual, for as yet the two birds are far from ready to mate.

At first the two bee-eaters frequently exchange roles. However, eventually a male begins to feel uncomfortable if he is repeatedly forced to assume a female role. This discomfort proves that he is a male. Moreover, the fact that the other bird repeatedly forces him to assume the female role means that it too is probably a male. Two females will also feel incompatible. At times a male and a female may feel uncomfortable together.

This shows that they are not compatible mates. A bird may go on "testing" marriage applicants for days on end. When two birds finally achieve perfect harmony, they spend the night nestling close together. The bee-eaters are now man and wife.

However, bee-eaters do not actually copulate until from ten days to two weeks later, when the two of them have dug their nest in the side of a cliff and "set up housekeeping." Thus for them, pair formation and copulation are two separate things. The fact that they become mates some time before sexual intercourse occurs proves that their "marriage" is based on the bonding instinct.

Ignorance as to whether one is male or female is not confined to the bird kingdom. Mammals too may pass through periods of confusion about their sexual identity. Sheep are reputed to be very stupid animals, and certainly bighorn sheep are no exception. These sheep live in herds in the Rocky Mountains. The only thing they understand is the distinction between superior and inferior, master and slave. Even during the act of mating, they never grasp the difference between male and female.

For most of the year, adult female sheep behave like one-year-old males. The size of their horns is approximately that of a one-year-old male, and the size of its horns determines the rank a sheep holds in its society. However, on the two days in the year when they are in heat, the females behave like rams and go around looking for trouble.

Soon the female in heat encounters another hotheaded sheep that wants to pick a fight. The two opponents rear up on their hind legs, race toward each other, and ram their horns together. The collision almost knocks them unconscious. Thus both sheep must wait for their heads to clear before continuing the fray. However, the two opponents behave very honorably. The battle cannot begin again until both of them are in position on their hind legs.

When the adversaries are evenly matched, the battle can continue for hours. However, sometimes it is over quickly. In any case, the loser stoops down before the winner and allows him (or her) to mount his (or her) back and to engage in

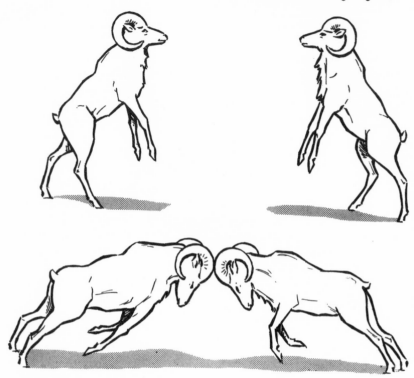

Duel between two bighorn sheep. The loser is treated as a female.

sexual intercourse. Female bighorns are the Valkyries of the animal world. No female sheep allows a male to mate with her until he has bested her in a fair fight.

If, as is often the case, the female defeats the male, he kneels down, she mounts him, and they pretend to copulate. More frequently two rams will fight, and in the end they perform a homosexual act.

A bighorn sheep society recognizes neither male nor female, but only victors and vanquished. Thus it is small wonder that bighorn sheep do not "marry" as bee-eaters do. Bighorns are not monogamous; nor do the males keep harems like elephant seals. The sheep practice sexual communalism, i.e., free love, mating indiscriminately with other members of their herd.

To be sure, the female is in heat only two days out of the year, and thus mates very seldom. Sexual communalism is rare in the animal kingdom, occurring only in species in which the females are in heat for brief periods. It almost appears that females can tolerate such an arrangement only if they very seldom mate.

The sexual relations of some animals are even more bizarre than those of the bighorn sheep. Species exist in which the males behave like females and the females like males—not just on occasion, but all the time.

The female phalarope, South American painted snipe, and button quail all wear the pants in the family. That is, she wears a brilliant coat of feathers, while the male must be content with "feminine" camouflage, plumage of a plain brownish-gray. Moreover, the females are larger and stronger than the males.

Of this bird group, the sexual relations of the American phalarope have been the most extensively investigated. American phalaropes winter in Argentina. In spring they settle beside the swamps and lakes of southern Canada and the northern United States. The females arrive first. Immediately they stake out breeding territories and defend them against other females the way males of other species defend their territories against males. Female phalaropes are black, russet, blue, and gray and have white stripes. When another female approaches her territory, the lady of the house exhibits threat behavior, strutting forward with an open beak. If this warning does not suffice, the two birds will actually fight.

A few days after the females, the males arrive at the breeding grounds. Unlike the females, they are docile and placid creatures which stand around passively while they are courted by the belligerent females. To keep the timid males from being afraid, the females repeatedly stretch themselves up to their full height and point their beaks at the sky. This is a reassuring gesture intended to show that the beak will not be used as a weapon and that the female means the male no harm.

At the conclusion of her ceremonial war dance, the Amazonian female suddenly begins to behave very humbly. Stooping down in front of the smaller male, she invites him to mount her and mate with her. To anyone familiar with the mating habits of other species, this is a rather grotesque sight. The female is so much larger than the male that even when she bends down, he cannot mount her directly. Fluttering his wings, he rises in the air like a small helicopter and lands on her back.

Incidentally, birds are quite capable of copulating if the female mounts the male instead of the male the female. Male birds have no penis. To mate, birds press their genital openings or cloacas together. (Cloacas are body openings which emit feces and urine as well as sperm or eggs.) They can do this just as easily if the female mounts the male as they can the other way around. Thus the sexual position they assume is a matter of personal taste.

Male American phalaropes are in charge of all the household duties, like lining the nest, sitting on the eggs, and caring for the young. Immediately after laying three or four eggs, the female departs to court another male. Thus the female phalarope is polyandrous.

What caused this reversal of the standard sexual roles?

Every living creature produces a variety of male and female sex hormones. Thus to some degree, every animal and every human being is a hermaphrodite. However, as a rule female hormones predominate in females and male hormones in males.

For the American phalarope, the gray phalarope, the button quail, and the South American painted snipe, this is not the case. The females of these bird species manufacture enough of the female hormones which stimulate egg production to lay eggs, and thus are nominally female. However, in other respects male hormones predominate. The males, on the other hand, are predominantly governed by female hormones.

The ovaries of the female American phalarope, gray phalarope, button quail, and South American painted snipe manufacture large quantities of male hormones. The hormones produce bright-colored plumage, bodies larger than those of the males, and masculine musculature. They also heighten aggression and make the female behave like a male. In other bird species, masculinity, beauty, and aggression go hand in hand. Here all three are attributes of the female. In addition, the pituitary gland of the male American phalarope produces large quantities of prolactin, the hormone which stimulates the mammary glands of mammals so that they can nurse their young. As a result, at breeding time the male loses many of his breast feathers. Blood rushes to the surface of his skin to keep the bare patches warm. The bare patches or "brooding spots" keep the eggs warm while the male is sitting on them and warm the young after they hatch. Because only the male produces quantities of prolactin, he must bear the responsibility of sitting on the eggs and caring for the young. In addition, his female hormones give the male a passive and peaceable disposition.

The sexual relations of these bird species reveal two things:

1. The more brilliantly colored one bird partner is as opposed to the other, the more aggressive it will be and the more careful it must be to curb displays of aggression toward its partner. The curbing of aggression takes the form of a courtship ritual in which normally aggressive behavior is converted into signals of friendship. The more beautiful the bird is, the less involved it is in building the nest, sitting on the eggs, and caring for the young. Moreover, beauty implies infidelity.

The reverse is also true. The plainer the form and coloration of the partner, the more humbly and peaceably it will behave toward its mate, and the smaller the role it plays in the courtship ritual. This rule applies to both males and females.

For example, when the female striped button quail, splendidly adorned in puffed-out feathers, dances like an American Indian chieftain to impress her mate, the plain-looking male must slowly and unobtrusively approach her and humbly lie down on his belly at her feet. In this species, the total capitulation of the male is a prerequisite of mating.

Only those bird species in which males and females look alike or very similar, conduct a courtship ritual in which both partners participate equally. Moreover, these pairs stay together for an extended period and build their nest and raise their young together.

2. The level of aggression determines the form of an ani-

mal marriage. If the males of a species behave aggressively toward other males and the females are peaceable, the males will tend to keep harems.

If both the males and the females of a species are aggressive, then the males will be hostile toward other males and the females toward other females. To mate, the two sexes must learn to curb their aggression toward each other. This can often be a dangerous and difficult process. By the laws of animal psychology, this kind of relationship results in monogamy.

If both males and females are docile, they will live in herds like wild goats.

Sometimes—for example, in the bird species I have just described—the females are very aggressive and the males docile. Such species practice polyandry, the female counterpart of the male harem. The barred button quail, which is native to India, represents an extreme example of the polyandrous animal society. The Amazonian females are so aggressive that the native population can capture them by tricking them into attacking artificial models of female barred button quails. The captured females are forced to fight as cocks fight in other countries.

Some readers may be wondering why I am spending so much time discussing bizarre animals hitherto known only to zoologists. What, they may ask, do all these creatures have to do with us? The answer is this: By studying extremes of behavior, we can learn to understand what constitutes more "normal" behavior.

The discussion of hormones and the fundamental nature of male and female plunges us into deep waters. However, perhaps my comments make it clearer why women are so often puzzled by the behavior of men and men by that of women.

At a time when the women's movement is helping to raise our consciousness of sexual roles, we may all be intrigued by the life style of bird species such as the American phalarope, the gray phalarope, the button quail, and the South American painted snipe. The females of these species have put into practice what many women can only dream of: freedom from household chores and the burdens of child-rearing. After courting the males and laying a few eggs, the female's task is ended.

The males must tend to building the nest, sitting on the eggs, and rearing the young.

At first glance it might appear that a polyandrous life style would help to perpetuate the species. The female can lay eggs in the nests of three, four, or five males. Moreover, instead of being totally useless or occasionally assisting in rearing the young, the males can tend their offspring while the females get on to the business of breeding more.

However, polyandrous societies are rare in the animal kingdom, for a polyandrous life style is not advantageous to a species. In the bird species I have mentioned, there is a large surplus of males and a relative shortage of females. Their drab coloration helps to camouflage the males, but many of the brightly colored females are killed by predators.

In the struggle of a species to survive, the loss of a female is a far more serious blow than the loss of a male. A female may be able to fill with eggs the nests of five males, but a male can mate with many more than five females. To perpetuate itself, a species needs a higher proportion of females than males, for each male can breed innumerable offspring, and therefore only a few males are needed to maintain an animal population. Thus it is disadvantageous to a species to practice polyandry. The few bird species which practice it survive not because of, but despite of, their unusual life style.

Few as are the species which practice polyandry, fewer still accord equal rights to male and female. The only such species I know belongs to the genus of rock partridges and lives in the chalky Apennine and Balkan Mountains. At breeding time, the female hollows out two nests in the thin soil with her feet. She digs the nests about one hundred yards apart. As soon as she has laid her eggs in the first nest, the male rock partridge takes over the nest and sits on the eggs. The hen assumes responsibility for the second nest. Thus each bird behaves as if it had its own children.

At first the monogamous rock partridges maintain separate households. To be sure, the male must leave to the female the labor of laying the eggs; but in other respects, they take equal responsibility for the care of the young. They maintain two nests one hundred yards apart. Thus if an enemy finds one nest, at least the other may be spared.

When the young in both nests have hatched, the parents occupy a single nest. For eleven months they raise all their chicks together.

Male and female rock partridges are very similar in appearance. Both measure some thirteen inches from head to tail; both weigh around fourteen ounces; and both have the same kind of protective coloration.

A rock partridge marriage seems to me ideal in every respect, and I can think of no reason why such unions are so rare in the animal kingdom. Equality between man and woman is equally desirable in human marriages—and equally rare.

# PART FIVE

# How to Win a Female

# CHAPTER 9

# Love's Laws of Harmony

*Sexual Partners Synchronize Their Emotions*

Doris was a female dove which led a melancholy existence in a zoologist's laboratory. For a year she lived alone in a glass cage, with nothing to do but look at the bare walls of the laboratory. She did not lay a single egg, for unlike domestic chickens, female ring doves do not lay eggs unless they have male companionship.[1]

In spring the zoologist placed next to Doris a second glass cage containing a male dove named Sam. The two birds could see each other through the glass, but they could not hear, smell, or touch each other. Despite this separation, within a week Doris had laid two eggs in an artificial nest inside her cage.

A female dove named Susie spent a year in isolation in an adjoining room. At the same time that Sam joined Doris, Potinus was placed next to Susie. Potinus was a castrated male, a eunuch. Unlike Doris, Susie did not lay any eggs.

How could the two females tell the difference between a

potent and a castrated male just by looking through the glass? Both females observed the behavior of the neighboring males. Potinus trotted languidly around his cage, interested in nothing but eating grain. But when Sam saw the female dove, he began to ecstatically court her. Puffing out his chest, he paraded up and down in front of her, periodically pausing to bow several times in succession, and executing elaborate dance steps. Unfortunately, the tiny cage somewhat cramped his style.

The mere sight of Sam performing his courtship dance was enough to stimulate the growth of Doris's ovaries and oviducts. Despite the fact that she did not actually mate with Sam, her sexual organs rapidly matured so that soon she was able to lay her eggs. Of course, all the eggs were sterile. Nevertheless, this experiment demonstrates that not the sexual act but the male's courtship stimulates production of the female's eggs.

The courtship dance of the male dove stimulates the secretion of the female's hormones. The hormones in turn stimulate the growth of the ovaries and oviducts until they are large enough to function. The whole process takes less than a week. Thus the "communications channel" that stimulates hormone production leads from the female's eye to an IRM in her central nervous system. If the eyes of a female dove are covered, she will never be able to reproduce by natural means. Of course, a zoologist can artificially introduce the necessary hormones into her body. Using this technique, he can induce the production of eggs at any time of year.

When male and female doves mate in the spring, the male's courtship dance serves to stimulate the growth of the female organs so that both birds are physically prepared to mate at the same time.

Most animals are able to reproduce for only a few days

The northern gannet is as handsome as he is aggressive. His beak is sharp and powerful, and he can inflict wounds resembling dagger-wounds on a human being. Fortunately his wife, which he beats every day, possesses a thick coat of feathers. Thus the male does not inflict fatal injuries on the female.

each year. The ovaries and oviducts of the female and the gonads of the male function for a very brief period. For reproduction to occur, the physical and emotional reactions of the partners must be carefully synchronized.

The process of synchronization takes place in a minimum of three steps. With each step, synchronization becomes more exact. However, before synchronization can occur, both partners must have reached sexual maturity. Many people assume that living creatures reach sexual maturity by a natural process of growth, i.e., simply by growing older and bigger. But this is not true of animals or of human beings.

Around 1870, European and American women experienced their first menstrual period somewhere between the ages of fifteen and eighteen years. Statistics show that one hundred years later, in 1970, the first menstrual period occurred between twelve and fifteen. Scientists attributed this swifter rate of maturation to various causes: the increasing size of human beings, the urban climate, improved nutrition, advances in medicine and hygiene, increasing immorality, and a score of others. All these explanations were purely hypothetical.

Then in 1971, J. G. Vandenbergh tried to solve the mystery by experimenting with mice.[2] His experiments indicate that hormonelike substances called pheromones may influence sexual maturation.

Normally a young female mouse is raised by her mother and associates only with her brothers and sisters. However, if a young female is raised in the company of sexually mature male mice, her first eggs will develop twenty days earlier than if she is raised by her mother alone. Moreover, if a female mouse is raised alone in a cage to which someone every day brings a little male-scented earth from the pen of adult males, the female becomes sexually mature twenty days earlier than is normal. Thus the male scent can induce sexual maturity as quickly as the presence of the males themselves.

The reverse is true of a female mouse which, while being raised by her mother, also associates with other adult females. The more time she spends with older females, the longer it takes the young female to reach sexual maturity.

The length of time involved in sexual maturation reflects

the current proportion of males to females in the population. Mice have developed a special form of population control. When there is a surplus of female mice, the young females take longer to mature and reproduce. If there are too few females, the young ones will mature rapidly to produce more.

Abundant protein and vitamins also hasten the sexual maturation of female mice. However, compared with social factors, protein and vitamins have an almost negligible effect.

Professor Vandenbergh believes that social factors which influence human females during childhood may be primarily responsible for the earlier sexual maturation of women in modern times. No doubt it would be an oversimplification to suggest that modern coeducation, the education of boys and girls in the same schools, effected the transformation. However, this or similar causes may well be involved.

For example, it is a fact that many young women who served in the women's work corps during the Hitler era and who lived in barracks for prolonged periods, ceased to menstruate. Scientists have yet to investigate the question of whether daughters raised in fatherless homes likewise experience an inhibition of the menstrual cycle. In any case, experiments conducted on mice have paved the way for further investigations of the sexual maturation of human beings.

Each year shortly before mating season, the sexual organs of many animal species mature so that the animals may reproduce. In order that male and female may mature simultaneously, an endogenous rhythm—i.e., a sort of "calendar" inside the organism which functions automatically—synchronizes their development.

The internal calendar is governed by the length of time the sun shines each day. Some species mate in spring, others in the fall. In spring the days grow longer. Around the beginning of March in central Europe, the sun will begin to shine for ten hours and fifty-four minutes a day. The days have now reached a critical length for larks. As soon as the sun shines longer than ten hours and fifty-four minutes, male and female larks begin to produce the hormones which stimulate the growth of their gonads, ovaries, and oviducts.

At the beginning of May, dormice awaken from their winter's hibernation. By the end of the month the sun begins to shine longer than fourteen hours and forty-six minutes a day. At this signal, the dormice's sexual organs begin to mature.

The reverse applies to species which mate in the fall. They receive the signal to secrete sex hormones when the days begin to grow short again. For example, on the day in mid-September when the sun shines for less than twelve hours and fourteen minutes, hormones begin to flow through the bloodstream of the red deer.

The internal calendar can be very precise. Thus all the members of certain species of migratory birds start their migration on the same day. However, sometimes the biological clock can run several days or weeks fast or slow. As a result, frequently male and female are not prepared to mate at the same time. Nature had to invent additional devices to synchronize the readiness of partners to mate.

One such device was to allow the males of some species to remain potent longer than the period in which the females are able to conceive. In harems, herds, and other highly organized animal societies, the males remain potent for prolonged periods. From time to time the pashas, or male harem-rulers, inspect the females in their territories and mate with those which happen to be in heat. Male birds that practice arena courtship are also potent for long periods. (See Chapter 14 on arena courtship.) However, in these species the female birds that are ready to mate select the males.

It may appear that this form of synchronization would function only in large animal societies, where there will always be some females in heat for the potent males to mate with. However, this is not necessarily the case. For example, does are far too aggressive toward other does to live in a large group like a harem or a herd, except during winter when food is very scarce. When people see a group of deer standing together, the group always consists of a doe with several of her half- or almost full-grown fawns. However, even among deer, the same kind of synchronization obtains. That is, the males remain potent for long periods.

During mating season, a doe which is ready to mate calls loudly through the forest, attracting wandering bucks which are searching for does. A hunter can imitate the doe's call by holding a beech leaf between his thumbs and blowing along the edge, causing the leaf to vibrate. If the hunter is skilled enough to put the right emotional intonation into his call, he can entice a love-drunk buck into his rifle sights.

No more exact synchronization is necessary to bring together the buck and the doe. However, in some species the males and the females are more aggressive than deer. In these species, the two sexes must be more precisely synchronized so that feelings of aggression and fear do not inhibit sexual attraction.

When they meet, an aggressive male and female are torn by conflicting emotions. If one of them displays aggression, the other will feel afraid and vice versa. Aggressive sensations can heighten sexual drive, but fear blocks desire and thus inhibits mating. Thus one or the other of the animals is not in the proper mood to mate. The two partners must harmonize their emotions by curbing their aggression and dispelling each other's fear. They achieve this through courtship ceremonial, which creates a feeling of sympathy between them.

Doves are not nearly as peaceful as people often assume them to be. The male dove's courtship dance serves to dispel the two partners' aggression and fear.

Courtship ceremonial has a threefold function. First, it enables members of a species to recognize each other. Second, it enables them to identify each other's (and sometimes their own) sex. Both these functions were discussed in earlier chapters. Finally, courtship ceremonial enables sexual partners to synchronize the development of their sexual organs so that they are able to reproduce.

Sometimes the best way to clarify the meaning of behavior patterns is to describe what happens when the behavior in question does *not* occur. For example, strange things happen when nature "forgets" to synchronize aggressive partners which are incapable of forming even a fleeting sympathy bond. We have already noted that tigers fall into this class, as do many creatures which developed early in the course of evolution, particularly snakes.

The sexual intercourse of snakes resembles a wrestling

match more than an embrace. Male and female recognize each other's sex by their smell. The two of them, and frequently one or more other snakes, tie their bodies in a knot known as a "Gorgon's head." Thus they imprison each other in the coils of love and at the same time imprison any rivals which happen to be on the scene. Zoologists do not know whether the female could escape the knot if she wanted to, or whether she is in effect being raped.

When the snakes are tightly knotted together, the male moves his tail toward the genital opening of the female and inserts his penis into her body. However, at this point the female's sexual organs are generally not yet fully mature; thus she is not ready to mate. Unlike that of doves, the sexual development of snake mates is not precisely synchronized. Thus initially the male fails in his attempt to mate with the female. The two snakes must stay locked together for hours or days until the female reaches the point of being able to conceive.

Snakes have no legs with which to grip rocks and the branches of trees. Thus one might assume that the snake lovers locked in their long embrace must engage in an artistic balancing act in order not to lose their grip on each other. In reality, the male snake need not fear that the female may slip from his grasp. When his penis is erect, it bristles with spines, bumps, and hooks. Once he gets it inside the female's body, it becomes firmly anchored and cannot be removed until the sperm has been released.

With the male's penis still attached to the female, the two snakes slowly unwind and lie apathetically side by side. Frequently the female, completely uninterested in what is happening, decides to travel somewhere else, and drags the smaller male along with her. If they encounter an enemy and the snakes attempt to flee through rough terrain, or if they begin to glide in different directions, the male's penis may simply break off. However, nature provided for this contingency by endowing him with two penises.

When male and female are not prepared to mate at the same time, a great many problems can arise. In such cases it is necessary for nature to take a few extra precautions. Thus the females of many snake species can retain active male sperm in their bodies for as long as five years, using part of the supply to fertilize themselves every year.

Now that we have seen what can happen when sexual partners are not carefully synchronized, let us return to the question of how males and females of various species succeed in achieving synchrony.

The behavior of doves reveals that synchronization can occur visually, i.e., when the female observes the courtship dance of the male. Other animals respond only to scents. If we take urine from an adult male mouse and dab it four times a day on the nose of an adult female, the female will be ready to mate one or two days later. However, she will mate only with the male whose urine she has smelled. If urine from a variety of different males is dabbed on her nose, the reverse will occur: Her sexual organs will go into regression. Thus for a female mouse, the scent of urine is not only an aphrodisiac, but actually guarantees fidelity.

A female mouse which has intercourse with several different male mice in rapid succession soon loses the ability to conceive young. Only females which remain for some time with the same male can reproduce.

During periods of overpopulation, mice begin to exhibit symptoms of degeneracy. For one thing, the females become promiscuous and continually mate with different males. Of course, they do not bear young. This behavior represents a form of population control.

Many other animals, particularly insects, employ scents as sexual stimulants. In my book *The Magic of the Senses,*[3] I extensively discussed the subject of scents used as sexual attractions, and in an earlier chapter of this book I mentioned the "love perfume" employed by the wood nymph and grayling butterflies. (See Chapter 4.)

In 1962, zoologists discovered a hitherto unknown method of synchronization, synchronization by taste.[4] A certain species of beetle lives among reeds along the shores of rivers. When a male accidentally encounters a female, he courts her by appealing to her sense of taste.

At first the female beetle is frightened and unwilling to mate. When the pair meet and feel each other with their antennae, she turns to run away. Then the male whirls around and presents his rear end, containing a special organ that produces something tasty to eat. Apparently the scent appeals to the female, for she promptly devours the aphrodisiac substance.

Finally the male beetle goes around to the rear of the female and feels her with his antennae to see if she is ready to mate. If she kicks him, it means "No!" Then the male goes on repeating the same ritual, sometimes for hours on end. The less mature the female's sexual organs are, the longer the courtship continues. However, eventually the chemical substance which the female has ingested matures her body until she is prepared to mate. The male waits patiently until the female does not kick him when he feels her rear with his antennae. Then he mounts her and mates with her.

Thus the way to this female beetle's heart is literally through her stomach.

Pain can also stimulate love. I have already mentioned the "love bite" administered to the females by male tigers and lions. Like Cupid, edible snails shoot small sharp arrows into each other's bodies to get each other in the proper mood for mating. Moreover, before mating many lizards bite each other in the throat and sides. Sometimes they even draw blood. If she chooses, the female lizard can flee and avoid being bitten. If she is willing to mate, she endures the pain.

The custom of inflicting love bites on mates began long ago in the history of evolution. Scientists have found the marks of lethal bites on the heads of dinosaur skeletons hundreds of millions of years old. The teethmarks show that these plant-eating dinosaurs were bitten not by carnivores, but by other plant-eaters. In prehistoric days, mating must have been an act of rape which frequently brought about the death of the female.

We can only guess that in the prehistoric world, animals had not yet developed courtship rituals, techniques of appeasing aggression, or signals to indicate friendly intentions. Thus the male dinosaurs may have had no choice but to attack the females, biting them until they submitted.

Lizards are the modern-day descendants of the dinosaurs. The males of almost all known lizard species still bite the females in the neck and sides. Moreover, the males of many carnivorous animals such as martens and skunks bite the necks of the females. These bites are relatively harmless.

Human beings may still experience a latent form of the drive to inflict pain on a sexual partner. Perhaps the behavioral and physiological roots of sadism and masochism lie in our prehistoric past, when love was still indistinguishable from pain.

Among human beings, love play frequently includes such behavior as biting, pinching, and scratching. This behavior cannot be viewed as pathological or overtly sadistic.

Frequently we express sadistic impulses in nonsexual areas of life, tormenting and tyrannizing over other human beings. Alexander Mitscherlich describes this phenomenon as follows: "Sadism plays a major role in authoritarian behavior. We rarely realize that a person who tries to dictate to others how they should behave, actually does so because he *enjoys* tormenting them. As a rule, we interpret such sadistic behavior as strength of character or a strong sense of moral values. In effect, we frequently idealize expressions of sadism. This makes it difficult for us to recognize and correct sadistic behavior in ourselves." [5]

Scientists are still debating the true causes of sadistic and masochistic behavior. In psychoanalytic theory, these phenomena derive from a sexual fixation of early childhood which causes a person to associate the ideas of love and pain. Sometimes psychoanalysts interpret sadism and masochism in terms of the aggressive drive. However, perhaps there is a third explanation. We may all have inherited from our remote animal ancestors a latent tendency toward sadistic or masochistic behavior. which can be activated by early childhood experiences.

Other forms of sexual perversion may also result from the reactivation of a dormant instinct which has been temporarily repressed or concealed by other instincts. Archaic forces seethe in us like magma in a volcano, waiting to erupt. Unhappy experiences or emotional shocks can bring these hidden forces to the surface.

Besides biting, animals may touch each other with arms,

legs, antennae, and other parts of the body to stimulate sexual desire. The bright red rock crab, which inhabits the strange lava cliffs on the coast of the Galapagos Islands in the Pacific, shows how important touch may be in stimulating desire.

Red rock crabs are cannibals. If a crab comes within eight inches of a somewhat smaller crab, it will leap on its back and cage it up between its eight legs. Then it will grasp the imprisoned crab with its claws and cut it into edible bites.

To protect themselves against their cannibalistic peers, red rock crabs form small societies, or "clubs," composed of crabs of the same size. Crabs are safe from other crabs of the same size. Moreover, these small societies offer protection against larger crabs.

Clearly it is very difficult for cannibals to mate. In one respect, the mating habits of red rock crabs are extremely unusual. This is one of the few species in which the male partner does the active courting and yet, during the sexual act, assumes a position *beneath* the female.

Swinging himself underneath her so that they are lying belly to belly, the male crab grips the female tightly so that she can neither run away nor attack him with her claws. Presumably the male lies on his back beneath her in order to curb his cannibalistic instincts, which might rise to the surface if he were to mount her back.

So that neither mate will eat the other while they are mating, the male red rock crab (shown in black) tightly grips the legs and jaws of the female. Then he swings himself under her body so that he will not be tempted to devour her.

Before the two crabs mate, they must conduct a dangerous and elaborate courtship ceremonial. In some unknown fashion, the male crab can tell from a distance of some sixteen inches away whether another crab is a male or a female. Naturally, he will only court a female which is approximately the same size as he is. Once he has found a female of his own size, he assumes a threatening posture. Raising himself on his two claws, he circles around performing "push-ups" the way male crabs do before fighting with other male crabs.

Then the male crawls toward the female. If she is unwilling to mate, she flees so swiftly that the male soon gives up the chase. However, if her sexual organs are sufficiently developed so that the two crabs have a good chance of achieving synchronization, the female does not flee. Instead, she slowly retreats, always keeping the male at a distance of eight inches. After a time, the male pauses to test the female's reaction. If she is willing to mate, she pauses too. When he sees that she has stopped retreating, he moves away, watching to see whether she will follow him.

If the female follows the male, the two crabs perform a lengthy dance, moving forward, pausing, and moving backward again. As if they were connected by an invisible pole, the dancing partners always remain exactly the same distance apart. For a long time they cannot bring themselves to move closer.

Finally the male halts when the female approaches him. Then he raises his two front pairs of legs as if he were saluting her, and the female begins to touch them. This part of the ceremony may continue for a long time. Frequently the courtship does not thrive. Either the female seems not to enjoy touching the male, or the male does not enjoy her touch. One or the other of them will then withdraw.

If the male begins to back away, the female may follow him. At this point, their sexual desire has largely subsided, and each fears the other because of the cannibalistic habits of their species. However, the male does not want to be thrown out of the "club" composed of crabs of his own size, so he stops running and presses himself flat against the ground. A few inches away, the female stops and does the same thing. Sometimes they may sit this way for half an hour without stirring. Then both of them move away in opposite directions, and the game is over.

Only if the female's touching of his legs sexually stimulates the male do the two of them mate.

Many insects, particularly members of the hundreds of thousands of fly, gnat, wasp, and beetle species, use a "touching code" that functions like an "Open Sesame." Only when a certain part of the body is touched will a sexual partner be willing to mate.

Before the female will mate with him, the male of one beetle species must beat out a certain rhythmic pattern with his right anterior leg on the third segment of the female's abdomen. If he fails to do so, she kicks him away. A male of another beetle species must touch a different segment of the female's body. Insects use countless recognition codes involving different rhythmic patterns and different methods of touching parts of the body with legs or antennae.

Thus many species of insects respond to the stimulation of erogenous zones. This kind of sexual stimulation—the triggering of an innate releasing mechanism by a key stimulus—achieves the synchronization of the sexual partners. Moreover, it prevents insects of different but similar species from making the mistake of cross-breeding.

Many species which are not members of the insect kingdom also engage in the stimulation of sexual partners by touch. For example, everyone who has visited a zoo has observed baboons picking lice from each other's fur. However, the primary purpose of grooming is not the removal of insects. Above all, the activity serves a social function. Stroking and scratching each other is a way for baboons to communicate affection and to establish and maintain social ties.

Once an animal behaviorist friend of mine and I went to an Amsterdam zoo to visit a champanzee which my friend had raised from infancy. The two of them were very fond of each other. After they exchanged greetings, my friend rolled up his sleeve and held out his arm. Tenderly the chimpanzee parted the non-existent fur on my friend's arms with its fingers and searched for invisible dandruff, lice, and ticks. Whenever it found an imaginary insect, it put the insect in its mouth and ate it. The chimpanzee was performing a ritual. This incident demonstrates how an action can become divorced from its original purpose and evolve into an abstract, symbolic gesture connoting closeness and affection.

Apes and monkeys appear to greatly enjoy being groomed and often close their eyes in bliss while another animal grooms them. We do not know whether they also become sexually aroused. In any case, they do not mate after grooming. Probably it gives them somewhat the same sensations that human beings experience while being massaged or manicured.

Zookeepers who tend apes and monkeys can usually tell when grooming is acting as a form of sexual stimulation. Suddenly animals such as baboons and Rhesus macaques begin to stroke particular parts of each other's bodies. This behavior indicates that two apes are about to mate. Thus apparently apes, like human beings, possess erogenous zones and know how to use them to promote sexual arousal.

In the wild, when two baboons want to mate, they leave the rest of the troop and go off alone into the bushes. Nonscientists may be tempted to interpret this behavior as a proof of modesty. However, as yet zoologists and animal behaviorists have discovered no proof that animals have a sense of modesty. The preference for privacy evinced by baboons and other herd animals such as elephants and ducks reflects their desire to escape the sexual envy of their peers.

Many animals cannot bear to see members of their own or other species mate. Sexual envy differs from the envy of another animal's food in that it does not imply that the envious animal wants the other's mate for himself. Even assuming that an envious male is physically prepared to mate, he would have to go through a long laborious courtship to win a mate away from his rival. As a rule, no animal wants to go to this much trouble. He simply wants to prevent the others from having any fun. Thus to escape harassment, the male and female baboon seek out a secluded spot.

However, in a nonsexual context I have observed animals

behave in a way suggesting that they may be capable of feeling embarrassed or ashamed. For example, if the highest-ranking rooster in a barnyard has been badly defeated in a battle with a younger rooster, he will hide for at least a half hour in the darkest corner of the shed. He stands there with his head hanging, facing the wall and not moving, like a child who is ashamed. He will go on standing there long after the victorious rooster is out of sight. Anyone who knows how distressed animals are when they "lose face" must admit the possibility that they feel something akin to shame. The rooster knows that when he comes out of the shed, he will have lost the respect of all the other chickens.

Once when my family and I were away from home all day, Senta, our female German shepherd, left her calling-card on the living-room carpet. When we got home, we looked for her everywhere. Finally we found her hiding in a dark corner under the baby crib. She cannot have been afraid of being punished, for we never struck her. I concluded that she must have felt ashamed. No doubt many dog-owners will agree with me that shame is not the exclusive prerogative of human beings.

In any case, the sense of shame cannot, as is currently the fashion, be attributed solely to the repressive effects of Christian tradition, bourgeois morality, and sexual inhibition. Chil-

---

Hundreds of thousands of guanay cormorants nest in colonies on the flat rocks along the Pacific coast of Chile. Yet despite the fact that each cormorant is surrounded by thousands of its fellows, it is primarily interested in its own mate. Guanay cormorants are monogamous. Their courtship consists of a series of ritual acts. First the male occupies a nesting site. Then, holding his head high, he beats his wings up and down, signalling to the female to approach him. Moving in a special flight pattern, the female signals her intention to land beside him. After she lands, the two birds continue the ritual by twisting their necks back and forth. At this point they are still free to reject each other. The female can simply fly away, or the male can drive her away. If they like each other, they run their beaks through each other's feathers like the two birds in the picture. Later the male goes out fishing. If the female is waiting for him when he returns to the nest, the two birds become mates.

dren in Israeli kibbutzim have been raised in complete sexual freedom, without being taught any rules about sexual behavior. Until around the age of ten, the children engage in intensive sexual play. Then they spontaneously begin to exhibit signs of sexual inhibition. They experience feelings of shame without really knowing what they are ashamed of. Thus the sense of shame is not, or at least not exclusively, something which we are taught to feel by other human beings.

Shame is an emotion. Thus we must be able to trace its roots back to the animal kingdom. The emotion of shame springs from the vast complex of emotions involved in the struggle for rank. More precisely, it springs from a sense of inferiority. In animal societies, the struggle for rank and the resultant feelings of inferiority are an everyday occurrence.

Human beings also experience feelings of inferiority, but their emotions are more complex than those of animals. A human being wants to amount to something in his own eyes. He forms an ideal image of himself and continually compares it with the reality, i.e., with his real self as he perceives it.

As a rule, our ideal image of ourselves involves the idea that we should be able to consciously control our behavior. When his instincts force him to commit acts he cannot control, a human being feels inferior and is ashamed. Alexander Mitscherlich describes the sense of shame in the following terms: "When a human being cannot live up to his ideal image of himself, he experiences shame and the fear of rejection and abandonment. The failure to live up to the demands of the Superego breeds feelings of guilt and the fear of punishment." [6]

Thus when they experience shame, human beings superimpose an intellectual valve judgment on an archaic, instinctive emotion. The sense of shame might be designated not as a conditioned reflex, but as a conditioned emotion. Our upbringing and current cultural values help to determine the causes for which we experience shame.

Sexual feelings arouse more shame than anything else. This may be because we view sexuality as the realm of the instinctual par excellence and do not like to admit that we are dominated by instinct. The desire to deny the role of instinct in their lives causes many people to ignore the true motivations of their behavior. They prefer to view behavior as resulting from conscious choice, for they are afraid that if they recognize the instinctual basis of their actions, they will have to feel ashamed of themselves.

Human beings also feel ashamed when they lie or commit acts they regard as wrong. Even if a liar does not outwardly reveal his shame, a lie detector, which measures his physical reactions, shows that his body has reacted to the lie. Accelerated heartbeat, elevated blood pressure, increased perspiration, and rapid breathing are all automatic, unconscious reactions which cannot be controlled by our will. The involuntary nature of these reactions indicates the instinctual character of shame. Moreover, such reactions reveal that a man possesses a sort of built-in "lie detector," which, functioning in communion with his emotions, tells him when he has done something of which he should be ashamed. This physiological phenomenon forms the emotional basis of conscience.

This brief sketch of the biological foundations of shame almost concludes my discussion of the methods of synchronization which prepare animal partners to mate. These methods include courtship dances, special scents and tastes that stimulate the growth of the sexual organs, the infliction of pain, and the stimulation of erogenous zones. Sounds may also play a role in stimulating the readiness to mate. Among these sounds are bird songs, frog concerts, and the drum-beating of drum fish. However, to elaborate on these examples would contribute nothing essentially new to the discussion.

I have yet to discuss modes of synchronization involving (1) the time of day or the weather, and (2) the establishment of emotional harmony between the two partners.

Not all times of the day are equally suitable for making love. Just as many animals are able to mate for only a few days or months in the year, so countless animals will mate only under cover of darkness, at dawn, or in the late afternoon.

As a rule, human beings are physically capable of making love at any time of day, but often they are not in the mood to do so. Every person obeys his own personal rhythm. Some people can hardly drag themselves out of bed and are irritable all morning, but become energetic and cheerful at night and do not go to sleep until very late. Other people are their best and most creative early in the morning and like to go to bed early. In the same way, different people experience sexual desire at different times of day. It is advisable that people give some thought to their personal biological rhythms before marriage, and make sure that their rhythms do not conflict with those of their partner.

Human beings are profoundly influenced by the weather as well as by the time of day. Different people can be stimulated or depressed by temperature and humidity changes, warm and cold winds, dry weather or perpetual rain.

The sexual drive of many animals is even more dependent on weather conditions than that of human beings. In his impressive courtship dance, a crested bustard in a Frankfurt zoo displays his feathers, makes knocking sounds, whistles, and leaps into the air. However, he ignores the onset of spring and is indifferent to the presence of a female bustard. He begins his dance only when the keeper turns on the spray hose in his aviary. He will also court the female when the keeper turns on the spray hose in the neighboring cage, even though he can only hear the water and cannot actually see it. The male bustard appears to be responding to a self-taught conditioned reflex.

What can cause a bird to respond neither to spring nor to a female, but only to the sound of rain?

The crested bustard is native to an extremely arid region, the Kalahari Desert of South Africa, where it rains at rare and irregular intervals. It would be fatal to the bustard offspring to be born at a time of drought. Thus crested bustards can mate only when rain begins to fall. And thus the bustard in the Frankfurt zoo performs his courtship dance only when he hears the sound of the spray hose. In the case of the crested bustard, the key stimulus which triggers the male's courtship comes not from the female, but from the world of nature.

The behavior of termites is even more uncanny. Termites are small creatures which live underground and virtually never emerge into the light of day. Thus most termites are blind. Only the males and females have eyes, which they use once in their lives, on the day they make their wedding flight.

For the wedding flight to be successful, the termites must meet the following conditions:

1. The termites of all the neighboring termite hills must rise into the air simultaneously so that the winged males and females of different termite colonies can meet in the air and mate without running the risk of inbreeding.

2. The flight must begin late in the evening so that it is still light enough for the termites to find each other, but dark enough so that their many enemies will have only a short time to find them.

3. It must not rain during the wedding flight. However, it must rain shortly thereafter, so that the ground will be soft enough for the termite mates to dig themselves a small shelter.

How do the little insects manage to fulfill all these requirements?

First, termites possess an internal calendar. The males and females, i.e., the future kings and queens, develop only during the rainy season. Moreover, termites have an internal clock which tells them when it is late evening above the ground. Finally, they possess an inner barometer which tells them when it is not raining but will very shortly begin to rain.

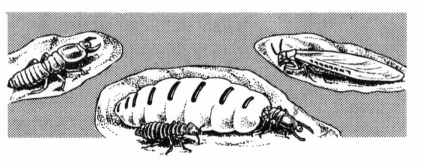

A termite society is made up of several types of termites. In the center of the picture is the large queen. Beside her is the small king, which always stays near her. On the left is a soldier with powerful mandibles. On the right is a winged male or female.

When all three of these key stimuli tell the termites that it is time to act, worker termites open the fortresses they have built at the top of the termite chimneys. Soldiers emerge in battle formation to see if the area is safe. If nothing disturbs them, they signal the other termites with a scent. Suddenly hundreds of thousands of winged males and females rise into the air, so that it looks as if pillars of smoke were rising from all the termite hills in the area. Thus termites of different colonies synchronize their activities with an almost uncanny precision.

By now it should be clear that love among the animals by no means represents a primitive and brutal phenomenon. The brief encounter of two sexual partners for the purpose of mating constitutes the most elementary unit of social organization. Yet even this brief encounter is a complex and problematic process involving highly refined techniques of courtship and synchronization.

The techniques of synchronization can be even more complex than those I have thus far described. Next we will examine how sexual partners synchronize their emotions before mating.

# CHAPTER 10

# Rules of Courtship

*From Duelling to Courtship*

Can a marriage last if the male beats up the female several times each day? It can, but only if the mates are a pair of northern gannets.

While the last storms of winter are still raging along the cliffs, the male gannets return to their hereditary breeding colonies on small islands or peninsulas in the Gulf of St. Lawrence and off the coasts of Newfoundland, Iceland, the British Isles, Brittany, and Norway. As soon as he arrives at the high, steep cliffs, a male gannet begins to drive away rivals from the nest which he and his mate occupied the previous year. If he cannot successfully defend his old nest, he will set about conquering a new nest that is still unoccupied.

Northern gannets are snow-white and about the size of geese. The males fight with a ferocity rare in the bird kingdom. Two males will grasp each other's heads and throats in their powerful beaks, and shake, twist, and jab at each other until both birds plunge off the dizzying cliffs. Recovering before they hit the water, they fly back to the disputed nest. During their brief absence, a third male has probably taken over the nest. The triumphant occupant greets both disputants with hoarse croaks and sharp jabs of his beak. Sometimes the battle may continue for two hours before one bird proves himself the un-disputed owner of a nest. Then he can enjoy a short "breather." However, soon another male looking for a nest will arrive at the colony, and the battle begins all over again.

For several days the owner of a nest does not dare to leave it. He cannot hunt for fish to eat, nor can he look for material with which to line his nest, which has been ravaged by winter storms. If he leaves the nest for even a moment, he will lose his home. He has no choice but to wait patiently until the squad-rons of females fly in from the mid-Atlantic and begin to circle around the colony.

To a human being, one gannet looks much like another. We cannot tell males and females apart. But even from a long dis-tance away, the females can recognize their mates of the pre-vious year. Unerringly each female picks out her mate from the mass of thousands of screaming gulls. Northern gannets, which have a lifespan of around twenty years, are monogamous, but mate only for the breeding season. For the rest of the year the males and females live apart. However, when the females arrive at the colony, they always remember their former mates.

At first, the male seems not to remember the female. He continues to behave aggressively toward anything that ap-proaches his nest and pecks at the female like a woodpecker attacking a tree. He also beats the female with his powerful wings. She has no choice but to endure the pain for several minutes without defending herself or striking back. In fact, the female even turns her sharp beak away from the male until he begins to calm down.

When the male's fury abates, the female turns her head toward him, and the two of them engage in a fencing match with their beaks. Gradually the playful exchange of blows be-comes a tender exchange of caresses. Billing and cooing, the two mates begin to lovingly run their beaks through each other's feathers in an avian version of "grooming." This is a classic case of the gradual transformation of a duel into an exchange of caresses.

No matter how savagely the male attacks her, the female does not retaliate. Gradually her docile behavior soothes him and changes his rage into love. In effect, she knows how to turn the other cheek—not so that it will be struck, but so that the male will lose the desire to attack her. If she were to show the slightest aggression, the marriage of the two birds would end in a flurry of torn feathers.

Normally the female gannet is far from weak or submissive. She is as large and strong as the male and is equally aggressive. Frequently the female guards the nest alone when the male is away hunting for fish, and she is quite capable of driving away enemy birds. Her mate is the only bird she treats with virtually Christian humility.

Why does submissive behavior have a soothing effect on aggressive animals? Instinctually based social behavior (social behavior involves the interaction of two or more individuals) constitutes a form of chain reaction. Responding to each other in turn, two individuals modify each other's reactions. For example, for a quarrel to occur, two human beings must first provoke each other, then insult and threaten each other, and finally come to blows. Each time one partner in the quarrel responds aggressively, his aggression triggers a more violent response in his opponent. If one of the partners interrupts the chain reaction by withdrawing or by pacifying the other, the quarrel will generally die down for lack of fuel. Submissive behavior eliminates the instinctual basis for continuing a quarrel.

In theory, human language and reasoning power should make it possible for human beings to calm each other's aggression and bring quarrels to a peaceful conclusion. Unfortunately, two factors frequently prevent their doing so. For one thing, human beings instinctively tend to counter aggression with aggression, thus escalating a quarrel. They rarely possess sufficient insight to break the vicious circle.

Second, if a person with insight tries to break the circle of aggression and offers his opponent the other cheek, the opponent instinctively tends to misinterpret the gesture as a sign of weakness and inferiority. Instead of respecting the other man's character and peaceful intentions, he despises him. The two adversaries may resolve the immediate quarrel peacefully; but the next time they fight, the battle will be more violent than it would have been if they had both given vent to their aggressive instincts the first time. On the next occasion, the more aggressive man will try to take advantage of the peacemaker and force him into submission. He now feels superior to his rival and assumes that if he pushes hard enough, he can make him give in. Eventually the more rational, peaceable man will be driven beyond the breaking point, and the resultant battle will be more devastating than any quarrel that had simply followed its natural course.

Rational and peaceable behavior is useless when one is dealing with sadists, disturbed personalities, mob violence, or repressive governments. Even Christ, who possessed to a superhuman degree the ability to soothe aggression, ended up being crucified.

Fortunately, northern gannets do not have to deal with such complex problems as human beings. The females pacify the aggression of the males with an efficiency that should make us envious. However, their wedded bliss does not persist very long. It takes the pair some twenty minutes to perform the ritual battle that ends in tender caresses. After that, the male leaves the nest to find food. When he returns, the battle begins all over again. The entire ritual must be repeated.

The married life of the female gannet seems like one long martyrdom. There is some evidence that her daily beatings may sexually arouse the female, for the male sometimes mates with her directly after a beating, dispensing with preliminary caresses. Nevertheless, hers is not a happy lot.

Only when a gannet couple grows quite old does the male cease to abuse the female every day. When one of the birds returns to the nest, they forego their combat and simply wave their beaks back and forth without actually touching each other. Now their battle is purely symbolical, a form of ritual greeting.

Northern gannets demonstrate how aggressive behavior can undergo stylization and function as courtship ritual. Moreover, they show how two animals which are initially enemies can mate and live together in relative harmony. The married life of many animal species is similar to that of gannets. However, most species omit the first act of the gannets' marital drama. That is, as a rule the male does not physically abuse the female. Most animals begin courtship on a symbolic level, engaging in a stylized ritual combat.

A male crane never harms his mate, but he often performs in front of her a dance recalling American Indian war dances. Like gannets, cranes are essentially solitary and belligerent creatures that do not enjoy the company of their own kind. If the species is to survive, somehow male and female cranes have to overcome their mutual antipathy.

Crane mates come to know each other far better than the male and female gannet. They are monogamous and mate for life. Cranes can live to be as much as fifty years old. Unlike gannet mates, a crane couple stay together all year long, even on the long winter journey south to Africa.

Throughout their long years together, the two cranes are torn between their impulse to attack each other and the need to curb their aggression. This conflict manifests itself in the male crane's dance. The dance is not exclusively a courtship dance but is performed throughout the year whenever the two partners come into close contact. Instead of beating the female every day, the male dances his way through their marriage.

The male crane leaps one or two yards into the air, spreading his wings, which have a total span of more than two yards, and pointing his daggerlike beak at the female. Konrad Lorenz has interpreted these gestures to mean something like, "Look at me. I am large, strong, and terrible!"

Then the male abruptly turns away from the female and tears up a clump of grass or tosses a branch high in the air. This behavior means: "To be sure, I am strong and terrible. However, my strength is directed not against you but against anyone who attacks us." In effect, the male crane is telling his mate that he can defend her.

He is not simply boasting. If another crane inadvertently approaches while foraging for food, the male crane will cease his war dance and move from words to deeds. Leaping in the air, he attacks the stranger, driving him away with savage jabs of his beak and flailing wings.

Meanwhile, the female observes the ferocity of her mate toward cranes other than herself. (As a rule, human females are also impressed by the male's display of strength toward other males and gentleness toward themselves.) Thus the male crane's dance serves to solidify the union with his mate.

In many animal species, the male's courtship dance before the female consists of threat behavior directed away from his mate. It is a form of courtship by threat.

However, when no intruder is present on whom the male can vent his wrath, his aggressive behavior can have disastrous consequences. The orange chromide fish, which inhabits the waters around southern India and Ceylon, needs a "whipping-boy" if he is to have a happy marriage. That is, he needs another fish which he can periodically thrash in front of his mate. (Of course, sometimes he is the one who gets thrashed.) If he finds no partner on whom to take out his aggression, he will eventually pick a quarrel with his mate. If this happens, the relationship between the mates is destroyed, and any young they may have or be going to have will perish. In a small aquarium where the female cannot escape the male's wrath, the battle will end in her death.

It is his lowering threshold of aggression that drives the male orange chromide to attack the female. (See Chapter 6 on instinct.)

A male orange chromide that is continually beaten in border disputes with neighboring fish reacts quite differently from the male that is deprived of a whipping-boy. That is, he does not react aggressively. After a number of painful defeats, he begins to wear himself out in "housework," digging one spawning hole after another. He will dig many more than are needed, and dig them much deeper than usual. If his neighbors have left him only a small territory to dig in, he will begin to fill in each hole before he begins the next. Thus the unhappy male engages in a never-ending labor of Sisyphus. Unfortunately, when he fills in holes that have already been filled with spawn, his young are destroyed.

The behavior of a defeated male chromide reveals how work can function to hold aggression in check. Every act consumes a certain quantum of energy. If a male animal exhausts his energy in mating or in physical labor, the energy is no longer available for aggressive acts—or at least not until he has rested and renewed his energy store.

Fish fanciers who want to keep a male and female orange chromide in the same aquarium must protect the female from the male's aggression. In a small aquarium, the male fish cannot occupy his time with digging. However, if a mirror is placed in the tank, it will draw off the male's surplus energy. The fish will attack his image in the mirror, thinking that it is another fish. He will return to the fray whenever he is feeling energetic and thus aggressive. No animal except the chimpanzee appears capable of plumbing the mystery of a mirror.

There is a difference between the marital relations of gannets and orange chromides. The female gannet seems resigned to her daily beatings. However, if the male orange chromide attacks her, the female fish regards this as grounds for divorce.

The difference in the behavior of these two species should warn us not to assume that all animals are the same or that animals can teach us how human beings ought to behave. For example, think how absurd it would be if we said: "Some animals can fly. Therefore human beings and guinea pigs also ought to be able to fly." Everyone will instantly recognize the fallacy of this statement and reply, "You cannot make such generalizations. Before drawing this conclusion, you would have to see whether guinea pigs and human beings possess the organs and the body structure necessary for flight."

By the same token, if animal behaviorists wish to avoid making fallacious comparisons, they must find out whether an animal possesses the instincts necessary to produce a certain kind of behavior. This is not always easy to do. After gathering his data, an animal behaviorist must frequently rely upon intuition to help him interpret it.

Some scientists study animal behavior for its own sake. However, the real importance of the discipline lies in the light it can shed on human behavior. Thus difficult as the task of interpretation may be, it is essential that animal behaviorists use what they have learned about animals to elucidate the behavior of human beings.

The difference between the behavior of female gannets and female orange chromides demonstrates the importance of levels of aggression in determining the character of a marriage. Gannets are so aggressive that they can mate only if the female allows the male to vent his aggression on her. If these birds were slightly more aggressive, or if the female refused to allow herself to be abused, the males and females could never come close enough to mate, and the species would not exist.

Orange chromides are somewhat less aggressive than northern gannets. Thus they do not engage in combat during courtship or after they mate, except when the male is deprived of his "whipping-boy." Gannets fight with their neighbors just as orange chromides do, but they are so aggressive that even the presence of whipping-boys does not keep the male from attacking the female.

How do these facts relate to human beings? Animal behavior reveals that levels of aggression largely determine the character of a marriage. Some human beings are much more aggressive than others. Heredity, upbringing, childhood experiences, and stress situations all help to determine how aggressive we are. The stability of a marriage depends on how the husband and wife respond to each other's expressions of aggression.

Secondly, the marital relations of orange chromides reveal the role that conflict with outsiders can play in a marriage. For example, if a man successfully overcomes many obstacles in his professional life and his wife supports his efforts, their marriage will probably be stable. (The role of the attractive secretary in the life of a successful executive is quite another matter.)

But what happens if the husband is unsuccessful in his profession?

Unfortunately, few unsuccessful husbands behave like the male orange chromide and go out into the garden to dig. Most behave like chromides that have found no opponents on whom to vent their wrath, and take out their aggression on their wives.

I have already noted that the reactions of one spouse—usually the wife—to the attacks of the other—usually the husband—largely determine the degree of stability in a marriage. Sometimes the disparity between the aggressive drives of husband and wife results in some strange combinations. For example, if a sadistic man marries a masochistic wife, they compliment each other's needs so perfectly that no strong sympathy bond is needed to hold their marriage together. He finds sexual satisfaction in tormenting her, and she enjoys being tormented.

The sympathy bond can persist for a lifetime. Sadomasochistic unions, on the other hand, may prove less durable. In such a marriage, neither partner values the other as a person. Instead, each values only the pleasure he or she obtains from the other. Under these conditions, either partner can easily find someone else to replace the other as soon as the marriage gets to be a habit and the pleasure it provides begins to diminish.

The rapist-murderer represents an extreme case of the sadistic-aggressive personality who treats human beings not as persons but as objects.

Some marriages are based on aggressiveness and held together by sexual attraction mingled with aggression and illusion. We have all met people who burst into sarcastic laughter whenever anyone mentions love. Once a woman told me, "All that talk of love is absolute drivel! My marriage is based on sex and nothing else but sex. My husband disgusts me. If it weren't for sex, I would leave him today—or maybe I'd kill him!"

It is disturbing to observe the pivotal role that aggression plays in every relationship between a male and a female. Among animals, mating patterns are primarily designed to solve the problem of aggression. The sympathy bond represents the most advanced technique of overcoming aggression. However, before playing her highest trump, nature devised many substitutes for the sympathy bond.

Before mating, most species must solve a twofold problem. First, when approaching a female the male must behave in such a way that the female does not flee from him in terror. At the same time, he must find some way of preventing her from attacking him. Thus the male must curb the female's instinct to flee as well as her aggressive drive. There are many possible solutions to these problems, of which I will discuss only three.

In South America lives a rodent called an akoushi that is somewhat larger and longer-legged than its close relative, the guinea pig. The akoushi has evolved a more elegant mode of courtship than another of its relatives, the agouti. The male agouti simply runs after the fleeing female until she drops from exhaustion. At that point, he is almost too tired to mate with her.

In his courtship, the male akoushi exhibits considerably more finesse. Female akoushis are capable of conceiving young for only a few hours. Thus the male must be able to locate and woo a female quickly. When he finds a female that is prepared to mate, the male begins to squeak plaintively like a hungry, helpless, cold little baby akoushi, thus arousing the female's maternal instinct.

Attracted by the sound, the female comes closer. Without stirring from the spot, the male begins to tremble violently. Thus instead of looking aggressive, he appears to be terrified. He does not dare to stray far from the female and thus cannot run a short distance away, as the females of many species do when they are being courted by a male. Instead, the male akoushi merely pantomimes flight by turning his back to the female and "running in place." As a rule, this combination of infantile behavior and apparent fear proves irresistible to the female akoushi.

Many animals employ similar tactics in their courtship ceremonial. I have already mentioned that human beings in love frequently behave like children. They may also exhibit real or simulated symptoms of anxiety such as apologies, blushing, stuttering, or turning pale.

No method of courtship is effective one hundred per cent of the time. Occasionally a female akoushi may remain unmoved by the male's timid and helpless behavior. In this case, the male abruptly changes his tactics. Standing up on his hind legs, he aims a stream of urine straight at the female a few yards away and soaks her to the skin, thus marking her as his property. Other males, especially males that are lower in rank, will not dare to court her. Since she is able to mate for only a few hours, she has no chance of finding another mate and will be forced to submit to the male that has marked her as his own.

The males of countless species, including human males, employ childlike or timid behavior to attract females. Thus clearly childlike behavior is a highly effective tactic during courtship.

Before he can woo a female, a male sparrow must have a nest. It does not matter if the nest is scraggly and untidy, for later the two mates can clean it up and make it more present-able. The important thing is that the male have some sort of nest. In a large sparrow community, many males do not own nests. Females will have nothing to do with these wandering gypsies.

Male sparrows without property frequently torment un-married females. Groups of them alight near a female and chirp loudly, courting her with widespread wings and tails held high. The female responds by assuming a threatening posture. Then the nestless suitors chirp even more loudly. Their choruses some-times attract the attention of human beings.

At one time it was assumed that these noisy courtships ended when the female consented to mate with one of the males. In reality, a female sparrow will never mate with a nestless male. Apparently the ineligible males harass females merely to express their indignation at their lack of property.

A male sparrow which owns a nest does not behave so crudely. When he encounters a female he admires, he chirps a few dissonant chords that a human being would find quite unappealing. To the female sparrow they sound like a tender love song. However, the female is primarily interested in prop-erty, not love songs. As soon as the male moves a short distance away, she will come to inspect the nest, ignoring her future bridegroom. Chances are that she will find it unsatisfactory and fly away. Thus the owner of the nest must do everything in his power to attract her. Spreading his wings, he trembles like an aspen leaf, bows down before the female, and thrusts his beak in the air, opening it wide like a baby sparrow begging its parents for food. This behavior arouses the female's maternal instinct and overcomes her natural antipathy to males.

Wolfgang Wickler points out that in their appearance, women resemble children more closely than men do.[1] This may be one reason why most men instinctively curb their aggression when dealing with women.

On other occasions than during courtship, animals may imitate the behavior of their young in order to curb the aggression of their peers and protect themselves from attack.

In addition to behaving like a baby sparrow, the male sparrow must demonstrate to the female that he owns property. People who think of birds as etheral creatures may be surprised to learn that for a female sparrow, material goods are the foundation of romance.

The female woodpecker is exclusively interested in the male's nest in the hollow of a tree. She tolerates the male only because of his property.

In the spring, the male greater spotted woodpecker drums loudly on a hollow tree, signalling other woodpeckers that he has finished carving out his nest. He drums not beside the nest itself, but in a nearby signalling tree. The signalling tree is hollow and resounds like an African jungle drum.

The woodpecker signals as if he were beating a war drum, trying to drive away potential rivals. Male woodpeckers rarely fight. However, when they do fight, they dig their claws into each other's bodies and hammer away as if they were hammering at a tree. The battle inevitably results in the death of at least one of the birds. Thus when the woodpecker beats his war drum, the blows symbolize the blows he will deal his rivals if they do not flee.

Except during mating season, the male's drumming frightens away female woodpeckers as well as other males. In spring the sound of the male's war drum still terrifies the female. It may take her as long as two months to overcome her fear and approach the male. Moreover, the male must struggle to overcome his fear of the female's aggression. The courtship may take several months. Thus frequently the male woodpecker begins to beat his drum in the winter, around mid-February.

For weeks the female continues to flee when she hears the male signal. When she flees, the male pursues her. Frequently it is not clear whether he is trying to drive her toward or away from his nest. There is even some doubt as to which of them is the aggressor. Thus woodpeckers engage in courtship by threat.

One day when the male is signalling, the female will fly over and cling to the same tree some two yards beneath him. Then the male performs a courtship flight. His performance recalls the graceful curves made by human ice-skaters or the hummingbird. Eventually he curves toward the tree where he has his nest. Perching beside the nest, he pecks enthusiastically at the trunk of the tree. He is showing the female her future home. If he cannot produce a nest, the female will not show the slightest interest in him.

The courtship ceremonial of the American red-bellied woodpecker is somewhat more elaborate. Like the male greater spotted woodpecker, the male red-bellied woodpecker sits beside the entrance to his nest and drums on the tree trunk to attract the attention of the female. Eventually the female flies over and clings to the trunk beside him. At this point the two birds are still strangers and thus are afraid of each other. To prevent the female from flying away, the male disappears into the nest hole and hides so that she cannot see him. Then the two birds drum a duet on opposite sides of the tree trunk.

Woodpecker mates never become close. Oskar Heinroth described their behavior as follows: "We observed that whenever one bird flew back to the nest, the other immediately flew away. We had the feeling that neither mate could endure the thought that another bird was helping it to raise and feed the young."[2]

The female woodpecker is devoted to the nest, not to the male. Many other female birds are just as "materialistic," including the female wheatear.

The female wheatear leaves to the male the task of finding among the rocks a hole suitable for use as a nest. Exploring holes among the rocks can be dangerous, for holes may already be occupied by lizards or snakes. A wheatear must find a place that is free of ants, for when the young birds hatch, the insects will strip the flesh from their bones. Moreover, the entrance

During courtship the male wheatear bird holds nesting material in his beak to signal the female that he wants to show her his nest.

must be so small that no cat can get through it; yet the hole itself must be spacious. Finding a hole that meets all these specifications is no small task. Thus the female hires a "real estate agent—i.e., a male wheatear—to do the looking for her.

Like a real estate agent, the male bird must advertise his property once he has found it. Hopping onto a large rock near the nest hole, he begins to sing loudly, displaying his charms by executing flights in the air, and occasionally landing on his rock to rest.

The male's singing and display attract the attention of the female. She remains some distance away, waiting to see what he has to offer in the way of property. Making no effort to approach her, the male busily hops from stone to stone, leading the female toward his nest.

When the male wheatear reaches the rock above the nest hole, he sings a special strophe, then hops to the entrance and bows twice. With his head bowed low and his tail spread wide, he enters the hole. Seconds later he emerges so that the female can enter and view his discovery. However, frequently the female will decide that she does not want to enter the hole. Perhaps she finds the entrance unprepossessing; perhaps, like any clever buyer, she does not want to appear too eager; or perhaps she is simply slow to get the point. The male seems to incline to the latter opinion. While the female perches indifferently on a nearby rock, he gathers a few blades of dry grass in his beak, hops back and forth in front of her, and then carries the grass into the hole to line the nest. This symbolic gesture is intended to convey two things to the female. First, the male is saying. "You dumb thing, I'm trying to show you our future nest!" Second, he is saying. "Look at me, I can build a nest!"

When the female wheatear finally consents to inspect the hole, the male hops up and down in front of the entrance, singing loudly.

During courtship, the males of many bird species show the females their nests and advertise their nest-building abilities.

Many species of birds build their nests on the ground, in a hollow in the middle of a field. One hollow will do as well as another. Thus the male has no specific location to show the female and must communicate his intentions in some other way. The South African quail finch is native to the African plains. During courtship, the male gathers clumps of dry grass in his beak and, holding himself erect, trots quickly over to the female. As he walks, he raises his feet very high as if he were marching.

Despite the fact that his mouth is full of grass, the male quail finch manages to chirp a seductive note when he reaches the female. Then he goes over to a spot chosen by him and begins to build a nest. He never finishes the nest, but that does not matter. What counts is the symbolic gesture.

Peter Kunkel kept two male and two female South African quail finches in a glass cage.[3] For some reason, neither of the females liked either of the males. Thus the birds never mated. Nevertheless, whenever the males, carrying nesting material in their beaks, proceeded to a chosen spot and began to build a nest, the females trotted along behind them. Soon the females turned around and left the nest. However, given their lack of sexual interest in the males, the fact that they followed them to the nest proves how attractive a male quail finch with grass in his beak must be to a female.

Male and female animals do not copulate until they have had time to get used to each other and get over their feelings of fear and aggression. In many species, pair formation is quite independent of the sexual act. Frequently two animals have been mates for some time before they engage in sexual intercourse.

Just before copulation occurs, South African quail finches behave differently than during the earlier phases of courtship. The male trots back and forth in a semicircle around the female. Each time he turns, he bows as if he were about to leap into the air. This gesture prefigures the moment when the male mounts the female. Here, for the first time, a sexual element appears in the quail finches' courtship ritual.

People who know relatively little about animal behavior and are unaware of the complex social problems involved in even the briefest sexual encounter, may read many sexual allusions into courtship ritual. In reality, sexual allusions during courtship are very rare. In the first phase of courtship, the two partners

These fish, known as kissing gouramis, join in a long kiss that expresses hatred as much as love. Protruding their thick lips, the two fish suck themselves fast to each other's mouths. Then they push and pull at each other until it is not clear whether they are trying to kill each other or are struggling to express their affection. Presumably the test of strength gradually develops into love. If a male and female are equally matched, their "kissing contest" will last a long time. Sometimes the combat continues for hours, and the two opponents may repeatedly separate and return to kiss again. If, after prolonged combat, neither fish is victorious, it is a sign that the pair are physically and emotionally compatible. Finally the kiss of wrath turns into a kiss of passion. Suddenly the two fish separate, mate, and then kiss again.

At times the infliction of pain can sexually arouse a female. A male sand lizard may bite a stubborn female in the throat, the flanks, or as in this picture, in the tail. Sometimes he bites her so hard that the wound bleeds.

must struggle to overcome their mutual fear and aggression. At this point, overt sexual advances appear to have a repellent or frightening effect on animals, much as sexual exhibitionism has on human beings. The ultimate purpose of the courtship dance is to bring about sexual intercourse; but overtly sexual behavior would make it impossible for male and female to get to know each other.

As a rule human beings, like animals, work up to sexual relationships gradually. (Clearly relationships with prostitutes are an exception to this rule.) Several years ago I saw a French comedy film which sheds some light on human courtship rituals.

Two experienced "ladies' men" and a shy, inhibited ascetic were travelling on a train. The two Casanovas wagered that by the end of the trip, their shy companion could not manage to kiss the pretty young lady in the neighboring compartment. The shy man, wanting to prove them wrong, wagered that he *could* kiss the young lady and joined her in her compartment. After that, he just sat there stiffly, staring out the window and not daring to say a word. When the train pulled into the station, he suddenly plucked up courage, seized the woman and tried to kiss her. Naturally she boxed his ears and screamed for help, and the supposed sex-offender was carted away by the police.

Meanwhile, one of the two lady-killers comforted the lady, carried her luggage, bought her flowers and chocolates, and took her home. At the end of the film, the shy man was languishing in jail while his rival was in bed with the lady. Thus in the sexual realm, the behavior of human beings is sometimes not far removed from that of animals.

# CHAPTER 11

# Gifts Seal Friendship

*Courtship Feeding*

At the end of the last chapter I mentioned that the Casanova in the French film wooed the young lady with gifts of chocolate and flowers. Gifts of food and other articles also play a role in many animal courtship rituals. Gift-giving among animals takes various forms. The male may seduce the female with a gift; he may give her something to eat so that she will not eat him during mating; he may bribe her into submitting to his advances; or he may use a gift to demonstrate his ability to feed her: "I can support a family!"

For some time, animal behaviorists debated whether male quails offered marriage gifts to the females. Then a rather malicious experiment told them the answer.

During mating season, a male quail was left alone in a large enclosure with a stuffed quail.[1] It was moving to observe the male ruffle his feathers and zealously court the inanimate female, trying to get her to respond. He tried to mount her no less than twenty-three times. Finally, in desperation, he fluttered over to his food bowl several yards away, speared a fat cater-pillar with his beak, returned to the female, and presented her with the tasty morsel. Thus a male quail will offer a female food only when all other measures have failed.

It is an amazing feat for one animal to offer another food. As a rule an animal must go to considerable trouble to procure food. Before giving it to the female, the male must overcome his desire to eat it himself. Moreover, he must be capable of understanding that the food will so enhance his charms in the eyes of the female that she will consent to mate with him.

The practice of courtship feeding did not develop as a result of the male's efforts to woo the female after all normal measures had failed. The roots of this behavior lie in the ability of the male to refrain from stealing food from his mate. The herring gull is one species which possesses this ability.

In the harbor of Wilhelmshaven, West Germany, Friedrich Goethe observed the behavior of a group of herring gulls.[2] It was December, long before breeding season. Some passers-by had thrown a roll to a female herring gull named Uta, which Goethe could easily recognize by sight. Picking up the roll in her beak, Uta flew away. At once a larger and stronger great black-backed gull set off in pursuit. Then Ulrich, Uta's mate of many years, flew shrieking across the waves and attacked the black-backed gull like a dive bomber. Driving away the would-be thief, he landed near his mate and allowed her to eat the roll, not attempting to get any of it for himself. If the two gulls had not been mates, he would not have been so gen-erous.

The herring gull does not actually feed his mate, but he demonstrates behavior antecedent to courtship feeding: He allows the female to eat without attempting to take food away from her.

At low tide, one often sees two herring gulls side by side on the sand. Two gulls that stay this close together are invari-ably mates. The male will allow the female to pluck a fat worm or a mussel from right under his nose.

Shortly before breeding season, the male gull not only

allows the female to eat without taking food away from her, but actually feeds her. Whenever he returns to their nest in the dunes, he brings her cockles. Then he behaves like a baby herring gull begging for food. In reality he is not asking for food, which he is carrying in his beak and crop, but for love. Here the gesture of begging has a symbolic meaning. When a male brings a female food and then begs for food, he is asking the female for her love.

Many dog-owners have observed their pets beg other dogs for love. A male dog which has been trained to sit up and beg for food may use the same trick to beg a female dog to mate with him. That is, he tries to use a gesture learned from human beings to communicate with another dog. Unfortunately, the female dog has no way of understanding what he means.

Many male insects instinctively practice courtship feeding. The male scorpion fly produces his own food gift, a quantity of tasty saliva. Scorpion flies have insatiable appetites. Thus unless the male gives the female something to eat while they are mating, she will eat him instead.

However, like male birds, male insects obey the precept, "Don't give anything away unless you have to." If the male scorpion fly encounters a female that is already busy devouring prey, he does not bother to give her his gift before mating with her.

The Oriental white-eye bird need not fear that the female will eat him. Nevertheless, she can get very angry. Thus the male keeps a supply of food on hand to feed her if they quarrel.

Bubi was a male Oriental white-eye which was trying to court the female Susi.[3] Bubi cautiously sidled along the branch where Susi was sitting until he sat by her side. She did not even favor him with a glance. Then he swung himself around the branch several times backwards, like a gymnast swinging around a horizontal bar. He concluded his performance by hanging upside-down from the branch with his breast directly beneath Susi. If she were impressed, she was supposed to respond by running her beak through his breast-feathers. The performer waited in vain for applause. Growing angry, Susi threatened the male bird with wide-open beak. Smoothly Bubi raised himself to the branch and thrust his beak into hers, inserting a morsel of food. As she choked down the food, Susi's wrath was appeased.

On another occasion, Susi was less angry and threatened the male with her beak closed. Bubi touched the tip of her beak with his and then slipped his tongue into her beak. It was not clear whether he actually gave her any food or was merely offering her a kiss symbolizing food.

The swordtail characin, a fish native to South America and distantly related to the piranha, does not have to keep a food supply to soothe the wrath of a quarrelsome mate. He possesses a special organ resembling food, with which he deceives the female.

The female swordtail characin is so aggressive toward males of her species that no male can mate with her unless he tricks her. When the male sees a female, he opens his gill cover and releases a long, almost invisible bonelike process. At the

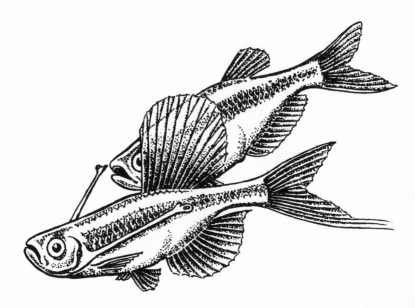

The male swordtail characin (below) "fishes" for a mate.

end of the process sits the imitation of a tiny crab. By jerking his gill cover, the male fish can make the tasty morsel "swim" through the water with the jerky movements typical of this species of crab. Waving the delicacy temptingly in front of the female, the male behaves as if he were about to flee. The female, believing that she has frightened him away, turns her attention to the supposed crab, darts over, and bites down on it. Now she is caught on the "hook."

Thus male swordtail characins actually go fishing for their mates. To be sure, the hook has no sharp points to hurt the female's mouth. The female simply chokes and keeps trying to swallow the bait. While she is busy, the male promptly mates with her.

Animals that live in groups do not have such acute problems in overcoming aggression.

Unlike northern gannets, whose breeding colonies are no more than large assemblies of enemies, terns are devoted neighbors. Groups of terns band together to defend their nesting grounds against enemies. Moreover, the birds have friendly relationships with terns outside their immediate families. In a nonaggressive society like this, the males do not need to seduce the females by offering them food. Thus here, offering food to a female has a different meaning. It is a way of testing whether or not the female likes a certain male.

When a male tern wants to mate, he catches an especially handsome fish and carries it back to the breeding colony. Here he marches proudly up and down in a group of eligible females, thrusting out his chest and holding his head high. As he struts, he inspects the females. Finally he offers one of them his fish.

The female must be careful not to accept the fish too quickly, for once she does so, she has pledged herself to be his mate. It is amusing to observe the female's reaction. She may ignore the male completely or even turn her back on him. She may look critically at the fish and the male and then walk on looking bored. She may even take the fish in her beak for a moment. However, she must give it back at once if she does not intend to mate with its owner. Apparently an especially handsome fish makes up for a lack of charm in the male bird himself.

If a male tern has pressed his suit on numerous females and been repeatedly turned down, he may offer the fish for sale to the entire colony, like a man trying to hawk his wares in a market place. If no female responds to the offer, he will eat the fish himself and then fly away to catch another that is more likely to win the favor of the ladies.

More frequently, one of the females will accept the fish. It is a sign of emotional harmony between male and female if the female shows no eagerness to eat the fish, but simply holds one end of it in her beak while the male holds the other. Sometimes the two birds may stand quietly side by side, holding the fish, for over an hour. He will not take the fish away from her and she will not take it from him. Among terns, this mutual generosity is the cornerstone of a happy marriage.

Terns are not the only birds that use food to test compatibility. The Bohemian waxwing conducts its test not with a fish, but with a berry, a fly, or the pupa of an ant. The gift is too small for both birds to hold in their mouths at once. Thus if they like each other, they will keep passing the tidbit back and forth: "Here, you go ahead and eat it." "No, you." "No, you." Neither bird ever actually eats the marriage gift. Rather than have just one of them eat it, they prefer to throw away the food they use in their symbolic ritual.

Cedar waxwings are slightly more materialistic. When they have mated they eat the food together.

As I have already mentioned, pair formation is asexual. Frequently two animals do not copulate until some time after mating.

The common or European bee-eater does not offer food to his mate during courtship but only before the sexual act. I discussed bee-eaters in the chapter concerning animals which do not know whether they are male or female. Offering food to another bird would not help a bee-eater to learn its sexual identity. Thus the male does not offer the female food until after they are mates. For from ten days to two weeks after mating, the two bee-eaters must work hard digging a nest in the hillside. They dig a corridor leading to the nest that may be more than two yards long. Only when the corridor and the nest are finished do the two birds actually copulate.

At this point, the female demands a gift. She will not stretch out her head, close her eyes, and crouch down to allow the male to mount her until he has given her a bee whose poison gland he has first squeezed and emptied. The bee is not merely a bribe. It also serves as proof that the male will be able to feed the female later on.

Once the eggs are laid, the two bee-eater parents take turns sitting on them, relieving each other at ten- to thirty-minute intervals. Thus the female has plenty of time to feed herself. Nevertheless, during the ten minutes or so that she is sitting on the eggs, she wants the male to feed her, and he does his best. After eating the insects the male brings her, the female coughs up their chitinous armor and uses it to cushion the nest. The two mates never bother to line the nest with feathers or wisps of hay. However, soon they are sitting on a soft bed of insect remains that bears witness to the male's diligence in bringing the female food. The male bee-eater's feeding of the female serves no biological purpose other than to line the nest with insect remains.

The American yellow-billed cuckoo feeds his mate not before, but during copulation. The Javanese coucal gives his mate food after the sexual act, as if he were paying a prostitute for her services.

The common European cuckoo practices polygamy. The male courts his current flame with nesting material, and as a mating gift, he presents her with an insect he has captured. However, the purpose of these gifts cannot be to demonstrate his nest-building talents or his ability to feed a family, for cuckoos do not build nests, sit on their eggs, or feed their young. Instead they lay their eggs in the nests of birds of other species and leave to strangers the task of hatching and rearing the young cuckoos.

Zoologists believe that at one time, the ancestors of the modern cuckoo laid their eggs in their own nests and raised their own young like other birds. At some point their breeding habits altered, but their courtship and mating patterns stayed the same. Thus the bringing of nesting material and insects to the female is a relic of the past serving only to signal her that she is being courted. The fact that this behavior has survived into the present demonstrates the ritual character of animal behavior during courtship and proves how readily such ritual behavior can become divorced from its original function.

Of the one hundred and twenty-eight existing species of cuckoos, only fifty species lay their eggs in other birds' nests. Thus the majority of modern cuckoo species hatch their own eggs and raise their young "like any decent, right-thinking bird." Some cuckoo species represent transitional forms between species that raise their own young and those that do not. Among these transitional forms is the yellow-billed cuckoo.

As a rule, female yellow-billed cuckoos build their own nests, but they do such a careless job of it that the nests are often destroyed during rainstorms. When her nest goes to pieces, the female will lay her eggs in the more solidly built nest of another bird. Then she does not simply fly away like the European cuckoo, but helps the strange female to sit on the eggs and raise the young. Neither mother behaves like a wicked stepmother, neglecting the young of the other bird. Instead, both mothers care for all the young.

Because the yellow-billed cuckoo does build a nest, it still makes sense for the male cuckoo to present the female with nesting material and food during their courtship.

Until a few years ago, zoologists were unaware that the males of so many different species practice the custom of bringing the females gifts. The Bohemian waxwing is content to offer his fiancée the symbolic gift of a single berry. However, the woodchat shrike, a close relative of the red-backed shrike, continues to bring the female food even after they have mated. Once a male woodchat shrike was observed to bring food to his mate twenty times in fifteen minutes. The female sat lazily on a branch, letting the male run himself ragged. Moreover, the male must have been hungry, for he brought every bit of food he found back to his mate. Despite his hunger and exhaustion, he went on feeding her until the time came for the birds to build their nest. From then on, both birds had to work so hard that the female could no longer enjoy the luxury of being waited on from morning till night.

The red-crested pochard inhabits the shallow waters along

the shores of the Lake of Constance. The male is sadly over-worked. He may have just recently found a mate or he may have already had one for more than a year. In either case, he must slave away from December until breeding time in the spring, bringing the female quantities of food to eat. Her feeding sessions may last for three-quarters of an hour, and she must be fed several times a day. The drake dives nine or ten feet underwater and pulls up aquatic plants by their roots to take to his mate. After nibbling a little, she throws most of the plants away, for the male keeps arriving with more. Even when she has eaten her fill, he keeps on bringing new loads to tempt her.

Zoologists believe that as breeding time approaches, the male's sexual drive is aroused before the female is prepared to mate. Thus the male must engage in hard labor to use up his sexual energy. The following incident seems to bear out this supposition: A male pochard named Bruno had spent ten minutes dragging up cartloads of aquatic plants from the bottom of the lake to give Berta, his "intended." However, Berta appeared totally uninterested and simply swam away. Bruno was so angry that, although he allowed Berta to depart unmolested, he promptly attacked and raped another female that just happened to be swimming by.

Macaws are among the most colorful birds in the world. Like female red-crested pochards, female macaws sometimes engage in orgies of eating. Unfortunately, in zoos a single macaw is often removed from its cage and chained outside on public display. A solitary macaw is apt to be very melancholy. In its natural habitat, the tropical rain forests of South America, the macaw is one of the most sociable of all living creatures. In some areas, these birds have almost become extinct; but where they do survive, they live in large, noisy flocks.

The macaw is one of the most tender lovers in the animal kingdom. Male and female mate for life and will help each other in times of trouble and defend each other against other macaws. Apparently the male believes that giving his mate an occasional gift helps to make for a happy marriage. He gives her the finest fruits he can find in the jungle and cracks hard Brazil nuts for her with his nutcracker beak. Interlocking their open beaks at right angles, the mates push tidbits back and forth into each other's mouths with their tongues. Then the male wraps his wings around the female and caresses her the way human beings caress each other with their hands and arms.

Macaws can live to be seventy years old. Thus frequently their marriages last longer than those of human beings. As the birds grow older, the feeding ceremony changes. Instead of actually feeding the female, the male simply performs the same gestures and caresses as if he were feeding her. Macaws practice the ritual of symbolic feeding, or billing.

However, as soon as the female has laid her eggs and is sitting on them in the nest, the male will begin to feed her again in earnest. Moreover, for some time after the eggs have hatched, the female does not leave the young birds. During this period, the male must feed the entire family. When he arrives back at the nest with food, he first gives it to the mother. She then feeds it to the young.

Many animals, like human children, seem to have a passion for ritual. The courtship of the blue-faced booby, a close relative of the northern gannet, involves an almost endless ritual in which the male offers the female a stone instead of food. In fact, the ceremony takes so long that he almost has to use a

Like bull gnus, bull impalas stake out and defend territories. However, a gnu is content with a little over three thousand square yards of African bush, whereas the large territory of a bull impala comprises between fifty and two hundred and twenty-five acres. Usually large neighboring property owners almost never catch sight of each other. Nevertheless, every day bull impalas meet several times to offer each other a ritual challenge. The picture shows two bulls challenging each other. If a bull impala sees a herd of females, he tries to drive them into the center of his territory. However, individual females or even large groups will bolt and run away whenever they get the chance. Thus in order to keep his harem, the male must be constantly on guard. As a result, he rarely has the leisure to enjoy his possessions.

Like their relatives the bull gnus, bull hartebeests try to plunge their sharp horns into each other's abdomens. To protect themselves from their adversary's horns, both animals sink to their knees before beginning their duel.

stone, for any food would probably have spoiled by the time he finished!

The male begins the ceremony by parading up and down in front of a group of females. His head and tail held high and his breast thrust out, he goose-steps arrogantly along the shore of the island where the boobies have their breeding grounds. As he walks, his webbed feet make the water splash up on the sand. When he decides that he has sufficiently impressed all the females, he struts over to the lady of his choice. bows, spreads his wings (he has a total wingspan of almost six feet), and whistles.

At first the female, apparently unimpressed, rejects his advances. Undaunted, the male struts around until he finds a stone the size of a chicken egg, picks it up in his beak, and lays it humbly at her feet. As a rule, she will not bother to give him or the stone a second glance. Then the male picks up his gift, trots over to another female, and lays it at her feet. One by one he will offer his "precious" stone to all the females in the area.

Sooner or later, one of the female boobies will show some interest in the stone. Picking it up in her beak, she turns her back on the male, waddles a few steps away, and drops the stone. Then the male picks it up, hurries after her, and offers it to her again. Once again she takes it in her beak, only to drop it a moment later. The game can continue for two hours before the female finally decides to accept the stone and the male along with it. Once she has made her decision, the two are mated for life.

I have already mentioned that in some courtship ceremonials, neither partner ever actually eats the food the male gives the female. A courtship gift is largely symbolic. This explains how the blue-faced booby can attach so much importance to an inedible stone. The booby has transferred to a stone the symbolic meaning that other species attribute to food.

The blue-faced booby is a close relative of the northern gannet; yet their courtship behavior is very different. The male blue-faced booby brings the female a gift to plead for her favor, whereas the male gannet continually abuses his mate.

Frequently, closely related species exhibit radically different behavior. One cause may be that levels of aggression vary markedly from species to species. Its level of aggression profoundly affects all the behavior patterns of a species.

In passing I have spoken of the "development" of social behavior. Only by analogy can we reconstruct the probable course of evolution hundreds of thousands or millions of years ago. It is difficult to tell exactly when certain behavior patterns developed. A change in the level of aggression, as well as other factors, may cause a relatively "modern" species to revert to archaic patterns of behavior.

The behavior of various species of dance flies reveals how the males of one insect family gradually made the transition from giving the females edible marriage-gifts to offering them "toys."

Dance flies are closely related to robber flies. Most of the 2,800 species of dance flies are murderous predators. The males, which are smaller than the females, have to protect themselves from their brides during mating. Before entering a swarm of dancing females, the males of several dance fly species capture a gnat or some other insect and hold it in front of them like a shield. When a female grasps the insect, male, female, and "shield" all topple to the ground. As soon as the female is busy sucking at her prey, the male mates with her.

Some species of dance flies have evolved from carnivores to vegetarians that suck the sap from plants. Only once in their lives do the females eat meat: when their suitors bring them an insect as a marriage gift. Males of these dance fly species are in no danger of being eaten by the females. Nevertheless, they have retained the mating customs of their ancestors. When a male gives a female an insect, she eats it even though she undoubtedly does not like the taste.

In North America live dance fly species in which the males bring the females gifts wrapped in beautiful packages. Using spinning glands at the tips of their forelegs, they spin fine snow-white silk threads around a captured insect. The males of some species give the females a fine large insect. Others, like pack-agers in modern supermarkets, have devised various methods of cheating their brides. For example, the male of the species

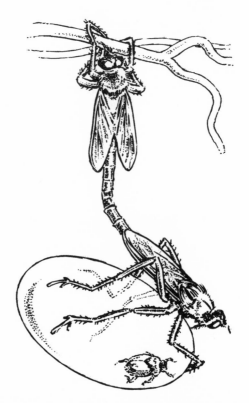

The female dance fly (*below*) is busy eating the prey her mate has given her. The prey is swathed in large quantities of "wrapping paper." Meanwhile the male (*above*), joining the rear of his body to hers, mates with her without fear of being eaten.

*Empis politea* spins a gigantic "blimp" about twice the size of the male himself to encase a very tiny insect. The female has to be content with this paltry gift.

The males of other species also wrap up small gifts in large packages. Still others play practical jokes on the females, giving them elaborate packages with nothing inside. Males of a species of dance fly called *Hilara sartor*, which is native to the Alps, do not deceive the females. Instead, they give them a purely symbolic gift. The male does not spin an empty balloon but a beautiful white veil, which he spreads out between his four middle and posterior legs. The veil in no way suggests food. When hundreds of the little flies, all carrying white veils, dance in the air with the sun shining on them, it is like watching elves or fairies dance. The glowing white of the veils attracts the females. Just as their less ethereal cousins accept the proffered gift of an insect, the females of this species accept bridal veils from the males and land in the grass. While the male mates with her, the female plays with her gift.

Bullfinch mates also feed one another, but not during courtship. For thirteen days after the female lays her eggs, the male and female take turns sitting on them. While the female is looking after the eggs, the male brings her food.

If the male bullfinch gets sick or begins to molt prematurely before the eggs are hatched, he "stays at home in bed," making himself useful by sitting on the eggs. Now the partners exchange roles, and the male begs the female for food just like a baby bullfinch. While he keeps the home fires burning, she goes out foraging for food to feed him and her young family.

If they are lucky, the female bullfinch will be able to nurse the male back to health. However, without the help of a veterinarian, sick birds usually die. But the female bullfinch's care will probably prolong the male's life. Frequently she can keep him alive long enough so that he can help her to hatch and care for the young birds. Thus at least their offspring will survive.

# CHAPTER 12

# Mass Courtship and the Marriage Market

*Group Courtship*

Animal love can take as many strange forms as love among human beings. Some animal species can only "get in the mood" for love by participating in a mass orgy. That is, they practice group or communal courtship.

Scene 1: Night, shortly after the full moon, on the Pacific coast of California. The spring tide breaks against the sandy beach. Swarms of fish resembling herring crowd toward the beach. Riding the waves, they land on the shore and begin to perform a mysterious dance.

Soon the beach is covered with fish wriggling like eels and sparkling like diamonds in the moonlight. They keep riding onto shore until the beach seems to be made of fish instead of sand. Then suddenly the hundreds of thousands of writhing bodies vanish like Cinderella's coach at the stroke of midnight.

From the end of February to the beginning of September, this spectral drama is repeated every two weeks during the spring tide, i.e., at the full and new moon. It takes place along a one-hundred-mile strip of coastline beginning northwest of Los Angeles and extending into the Mexican-owned peninsula of Baja California. Not many years ago, people believed that the legendary "dancing fishes" or "sea elves" were attempting to commit mass suicide.

The actor in the drama, the ecstatically dancing "sea elf," is a silvery fish called the grunion, which reaches a length of six inches. It dances on the beach only at night. A close relative, the sardine silverside, dances on northern coasts only during the day. Many of these daylight dancers are devoured by birds and other predators.

Do the swarms of grunions succumb to mass psychosis, swimming to shore in panic as small-toothed whales sometimes do? Or are they really bent on self-destruction?

No animal ever intentionally commits suicide. A dog whose master has died may stop eating and die of starvation, or an ibex pursued by hunters may plunge over a cliff; but neither animal intends to commit suicide. The dog does not choose to die. He has merely lost the will to do anything, including eat. The ibex, torn between the fear of going over the cliff and the fear of the hunters, chooses what appears to be the lesser of two evils. To commit suicide, one must be consciously aware of the meaning of life and death. Only human beings and chimpanzees understand these concepts. (See my book *The Friendly Beast*.[1])

When psychoanalysts talk about a drive toward self-destruction in human beings, they are not speaking of an instinct we have inherited from our animal ancestors. Instead, this drive represents a response to conditioned emotions. (See Chapter 5 on brainwashing.)

Thus the grunions' exodus from the sea onto the land does not represent a form of mass suicide. The fish are simply burying their spawn in the sand, for the eggs would die in the water.

Shortly before the grunions stream onto the beach to spawn, they gather in the shallow water at the crest of the high tide, waiting for the tide to turn so that they can begin their courtship dance. When the tide begins to ebb, a few grunions jump onto the beach. Presumably they are a "scout troop" sent ahead to reconnoiter. If they do not return to the water, the other grunions will not come on shore.

Sometimes tourists or fishermen capture the "scout" fish. If this happens, the other fish may postpone their visit to shore for another two weeks, until the next spring tide.

Once a group of tourists lit hundreds of fires along the shore, planning to cook and eat the grunions as soon as they emerged from the water. However, not a single fish appeared, for the light of the fires kept them from coming to shore to spawn.

Only if the fish feel that they will not be disturbed will they writhe onto shore. Then they begin a race with time. Unlike eels, grunions cannot breathe outside the water, and they can only "hold their breath" for two or three minutes. In this brief time, the females must make their way to the point on shore where the sand has been dampened by the highest waves of the high tide. When a female reaches this point, she twists back and forth, boring her tail into the ground until she is

Grunions hold a mass courtship on the beach at night.

standing in a vertical position with one third of her body buried in the sand.

Then two, three, or more males writhe around the female, fertilizing her one to three thousand eggs the moment she releases them. All the fish race to complete their mass spawning before their time on land runs out.

Presumably the stress involved in moving on land and going without air sexually arouses the male and female grunions, synchronizing them so that they are simultaneously ready to mate. Then as soon as the eggs are fertilized, the thousands of fish vanish into the ocean whence they came. Four, six, or eight weeks later, the drama begins again.

The apparently chaotic mass of writhing fish perform their task with astounding precision. If they spawn too soon, the rising tide will wash the eggs out of the sand and destroy them. If they spawn too late, they will not be able to lay the eggs high enough on shore, and twelve hours later the next tide will wash them away. In order that the spawn may develop undisturbed for at least one week, the grunions must mate when the spring tide is at its full.

For years people have wondered how grunions can know more about high and low tides than any human being can know without a tide-table. Grunions must possess an inner biological clock governed by both the sun and the moon, which can "calculate" the relationship between them to determine the time of the spring tide.

After spending eight days in the warm sand, the young fish inside the eggs are ready to hatch. But they stay inside their shells—a sort of "waiting room"—for another week, until slight movements of the ground tell them that the next spring tide is flowing. Then, within a period of three minutes, all the young fish slip out of their shells and let the waves carry them out to sea.

If the winds blowing offshore keep the spring tide from rising high enough to reach them, the young fish wait for another two weeks inside their shells. When the next spring tide arrives, they awaken from their sleep like Sleeping Beauty waking to a kiss.

Strangely enough, the eggs of the grunion die in water. Thus the fish must spawn high up on shore. However, other species engage in similar mass orgies for no visible external cause.

Scene 2: East Africa. Against the deep aquamarine blue background of Lake Natron shine the magical pink shapes of eight hundred thousand flamingoes. It is almost twilight. All day the graceful birds have been filtering small insects from the knee-deep water. Now they have had enough to eat and are standing around idly in the lake.

Suddenly one of the flamingoes raises its head and, pointing its beak straight at the sky, utters a prolonged cry that resounds deep in its throat. Then the bird spreads its wings wide, revealing the black feathers on the underside. With wings spread and beak pointing upward, the bird begins to march proudly through the masses of flamingoes. Suddenly it jerks its head to the left, then just as abruptly turns it to the right.

The bird continues its stately parade. Before long other birds point their beaks at the sky, spread their wings, and begin to march. The guttural sound in their throats grows louder. The marching males and females attract other birds that want to join the parade. More and more flamingoes swell the ranks. A vast squadron cuts a path through the water, leaving behind the birds that do not care to dance. Finally hundreds or thousands of birds are swaying together, each bird moving its head to a different rhythm. Pink, white, and black feathers flash, wings flap, flamingo feet stir up the water, and the hoarse voices croak in ecstasy.

This scene by no means represents the overture to an orgy of group sex. The purpose of the ceremony is to separate from the rest of the colony birds that are almost ready to mate, and to gather them into a group. The birds that are not yet physically prepared to mate feel no desire to participate in the dance. Thus the dance serves to synchronize a whole group of flamingoes instead of individual pairs.

A few days later, the birds in the group that danced together pair off and mate. The couples do not mate simultaneously, but each in their own time.

Scene 3: The middle of December. The Lake of Constance is almost completely frozen over. The sun is shining, but the temperature is below freezing. At one point, the site of an underground spring, a patch of water is still free from ice. Various species of birds—swans, gulls, and ducks—are romping in the water. Despite the cold, the ducks have already opened their "marriage market."

In one spot, three male northern mallards are about to perform a ceremony with rules more stringent and complex than those governing a levée at the court of King Louis XIV. One after another, the drake must execute the following movements: shake his tail, rock his whole body up and down, shake his tail, swim while nodding his head, dip his bill into the water, toss a few drops of water in the air, grunt and whistle, shake his tail, assume an imposing posture with his wings folded together above his back like a napkin, gaze at an imaginary female, swim while nodding his head, turn the back of his head toward the female. Then he must repeat the same series of movements again, and then again, and so forth.

After the three drakes have repeated the same exercises six or seven times, two females arrive to watch them. The males concentrate on their performance like ice-skaters practicing for a competition. Then three more drakes arrive squawking "rabe-rabe," and begin to perform like the others. Two more females join the onlookers. Finally some seventeen drakes are dancing before an approximately equal number of females, which have formed a circle around the dancers. All the females are looking for their future mates.

To impress his female public, a drake must give a perfect performance, not omitting a single dance step or making a single mistake. If a drake proves himself a real virtuoso, one or more of the females, carried away by enthusiasm, may swim through the group of males nodding their heads vigorously up and down. This is the females' way of applauding. Their applause may stimulate sloppy performers to improve their act.

During the performance, male and female mallards begin to establish a relationship. A female begins to watch one male with particular interest. He responds by tossing drops of water in her direction and looking at her when he comes to the part of his dance that demands that he look at a female.

After the dance, the male and female ducks swim away together. In December, their gonads and ovaries have not yet developed. Thus it is impossible for the birds to copulate. For the time being, they are simply engaged. If, after spending several days together, they find that they are not compatible, they break their engagement, attend another dance performance, and both look for new partners.

Months later spring arrives. Now that their sexual organs have matured, it is time for the birds to mate. As a rule, duck couples remain faithful for life. Ducks which already have a mate do not visit the annual marriage market.

The market spares bachelor drakes the trouble of scouring the lake country looking for potential mates. All the participants and the onlookers know from the beginning why they are there. This form of group courtship enables male and female to "get down to business" without going to all the trouble of soothing each other's aggression and winning each other's trust. Under these circumstances, courtship simply represents a test of mutual sympathy and compatibility.

# PART SIX

# Animals Incapable of Marriage

# CHAPTER 13

# Without Power There Is No Sexuality

*Territorial Behavior*

Standing on a hill, you can look out over the grassy plains in the hollow of the Ngorongoro Crater. Twelve miles across, with circular walls almost two thousand feet high, the Ngorongoro is the second largest volcanic crater on earth. It forms a "natural zoo" containing one of the finest collections of animals in East Africa.

As far as the eye can see, at one-hundred-and-fifty-foot intervals all across the plain, stand the dark figures of bull gnus, their heads held high, posing like monuments in the landscape. There may be as many as two thousand males, all awaiting the arrival of the female herds. Every bull gnu possesses a territory of around three thousand square yards which he defends against other bulls.

Wildebeests or gnus resemble woodpeckers and sparrows in that only those males that own property have any chance of mating and reproducing. The same is true of many species of gazelle and antelope.

Irritating as this fact may be to ultra-leftist political sympathizers, the devotion to private property is not a trait we humans invented ourselves, but one we inherited from our animal ancestors. It was not, as Jean-Jacques Rousseau believed, the corruption of civilization that bred the "unnatural" urge to own property. On the contrary, we are all born with an instinct to acquire and defend territory. As inhabitants of Communist nations can testify, the official espousal of Communist doctrine does not necessarily abolish the instinctive desire to own property.

The life of the wildebeest shows how the ownership of territory can affect the aggressiveness of an animal, its sexual drive, and its entire social behavior. Because of the crucial importance of territorial behavior, I will discuss the wildebeest at some length.

At the age of between one and three years, male brindled gnus live in bachelor herds of fifty to five hundred animals. Some years ago, observers believed these male "clubs" to be social paradises, for the bachelors coexist without any form of social organization or ranking order, and they almost never fight. Thus in the bachelor herds, no weak animal is ever oppressed by the strong. However, although the young bachelors live together, they are not really close. If hyenas attack the herd, the gnus, unlike zebras, do not band together to drive away their common foe.

When a herd of fifty to five hundred bachelors enters the territory of a bull gnu, they all lower their heads and make a wide detour around the property-owner. If he attacks them, they all run away at a slow gallop. They behave this way even though every one of the herd is approximately as strong as the bull. Moreover, none of the bachelors has the slightest chance of mating with a female.

It would be a mistake to describe the herds of young bachelor males as social units. Instead, they are a sort of asocial aggregation of apathetic individuals. The herds constitute "open societies." That is, none of the bachelors has any objection if a new member wants to join the herd; and if an old member disappears or is eaten by a lion, none of the other gnus gives him a second thought.

In 1969, Richard D. Estes refuted the view that the herds of bachelor male wildebeest represent a "social paradise."[1] He showed that membership in a bachelor herd renders a wildebeest incapable of resisting property-owning bulls. Moreover, the bachelors allow a few tyrants to drive them away from the lush meadows into regions where the grass is so tough as to be almost indigestible. This tough grass provides excellent cover for lions which choose to stalk the herd. The spiritless bachelors are completely impotent. Their impotence derives not from physical but from emotional causes.

As soon as one of the young bachelors acquires property, he immediately ceases to behave like a pariah, displays a strong sexual drive, and becomes very aggressive. Instead of standing around humbly, he raises his horns and stands as if he were posing for a sculptor. Sometimes he gallops around his territory with a rocking-horse gait so that every other gnu can tell from far away "Aha, there is a landowner." The proud bull continually quarrels with his neighbors, drives away intruders, and often mates with several females in a row. Thus a gnu's character can alter so radically in the space of a few hours that one can scarcely believe one is looking at the same animal.

Sometimes a bull that owns territory gives up his property or is driven away. If this happens, his aggressiveness and his sexual drive instantaneously vanish. Docile and resigned, the ex-bull trots over to a bachelor herd and henceforth shows no interest in anything but grass.

Thus the possession or loss of territory has a profound emotional effect on a gnu. Being master of a small strip of grassy plain gives a male gnu a feeling of self-confidence and superiority to other gnus. When a bachelor acquires property, his inferiority complex disappears, allowing his natural aggressiveness and sexuality to reassert themselves. If he loses his land, the result is "psychological castration."

Thus the ownership of territory is mostly a state of mind. Without territory, no male gnu can reproduce.

Many species of animals permit only a few "elect" members to engage in sexual relations. For example, the "cattle system" is widespread. In the wild, only the top-ranking bull in a herd of cattle is allowed to mate with the cows. He prevents all his weaker rivals from approaching the females. However, he does permit several "psychologically castrated" lower-ranking bulls to remain in the herd. He seems to need the presence of a few inferiors to give himself a sense of superiority.

The Asiatic water buffalo completely loses his sexual potency unless he has one or two weaker bulls in his harem to bully whenever he chooses. Thus the mere possession of a harem does not satisfy his emotional needs. To function, he needs the presence of several emotional eunuchs. However, the "eunuchs" do not lose their physical potency. As soon as the "boss" bull dies or becomes ill, the second-ranking bull is quite capable of assuming all his functions.

The inhibition of aggression also inhibits sexuality. In cattle and gnus, sexuality and aggression are two sides of the same coin. However, cattle and gnus have different methods of acquiring the sense of power that stimulates their sexuality. The highest-ranking male bull in a herd of cattle automatically acquires sexual rights to the females. But neither the gnus in the bachelor herds nor the property-owning bull gnus have any ranking system. Every property-owning gnu has absolute authority over his own territory. Thus instead of a ranking system, gnus have a two-class system. Membership in the upper class confers on a male gnu a sense of power which stimulates him sexually. A male gnu belonging to the lower class remains sexually inhibited.

In 1967, around 12,500 gnus were living in the Ngorongoro Crater. Of this number, 3,300 were adult bulls. At mating season, some 1,900 of the adult bulls possessed their own territories. The other males were forced to remain bachelors, for there was no land left for them to claim. However, the bachelors always hang around "in reserve," waiting to take over a territory if one of the upper class dies.

A comparison of gnu and cattle societies reveals that high rank and the ownership of territory are important not in themselves, but for the sense of power they confer. It is the sense of power which releases aggression and breaks down sexual inhibition. Perhaps the sense of power serves much the same function in human beings.

Bull gnus love their territories more than they love female gnus. When they fight, they fight over territory, not females. There is no reason to fight over females, for no bull can really own a female. All the bulls must wait quietly inside their territories until a herd of cows condescends to visit them.

Bull gnus react quite unpredictably to the presence of females. A bull may receive a herd of cows with warm interest or with indifference; or he may even drive them away. His reactions depend on his personal temperament, how tired he is, how many females there are in the herd, what season of year it is, what the weather is like, and a score of other factors.

Even if a bull allows a herd of cows to enter his territory, he may be more interested in scuffling with his neighbors than in entertaining his guests. He will dart back and forth, poking one neighbor in the face with his horns and pushing another in the side until his female visitors, growing bored, slowly move off in a column toward another territory. Seeing the females depart, the bull may dash after them, attempt to block their path before they leave his borders, and harry them like a sheep-dog trying to round up his sheep. Despite his efforts, the females will probably keep right on walking calmly into the next territory.

Bull gnus behave very awkwardly around the cows, for apart from the few moments when they mate, the males and females never associate with each other. Apparently the cows are most attracted to bulls which quietly mate with them without making any kind of fuss.

As soon as they have mated, males and females lose all interest in each other, and the cows continue on their way. Male and female gnus do not feel any kind of sympathy bond and thus are incapable of pair bonding or "marriage." The females do not even constitute a harem, or if so, then only for an hour's duration. Gnus are asocial creatures which cannot develop highly structured social orders. In fact, when the cows enter his territory, the bull gnu behaves in a highly antisocial manner, driving away all the young calves accompanying their mothers. He drives away young females as well as males. Thus it is impossible for gnus to form a complex social unit.

The examination of complex animal societies—schools of fish, breeding colonies, insect societies, hoofed animal herds, harems, or random groups created by an abundant food supply or by mass reproduction—reveals that the core of all social units is the family. This is true of ant, bee, and termite societies as well as of wolf packs, schools of dolphins, and troops of apes.

To be sure, as an animal grows older, a larger social unit may become more important than the family. For example, this is the case with some ape species. Nevertheless, all complex social orders in the animal kingdom ultimately derive from the family unit.

Many factors tend to prevent a species from developing family units. I have described the difficulties a male and female must overcome before they can engage in even the most fleeting sexual contact. Above all, they must somehow curb their aggressive impulses. The second step in the development of a family unit is the overcoming of aggression toward offspring. In a future book I will discuss how parent animals achieve this. Third, to create a family unit, an animal must overcome its aggression toward its young after they are adults.

Gnus are incapable of developing a family unit, for they cannot fulfill the third of these requirements. Their hostility toward their adult young prevents them from creating an elaborate social structure.

Even when no females are present, bull gnus seem totally devoted to their few square yards of African plain. If lions or human beings drive a bull away from his territory, he will always return to the same spot. During the dry season, when no females visit them for months on end, some bulls leave their territories and roam with the bachelor herds. But at the beginning of mating season, they are back on their home ground. Just as a migratory bird returns each spring to the tree or bush where it built its nest the previous year, a bull gnu unerringly picks out his own territory on the monotonous plains.

Until females arrive, a bull gnu has only two things to do: eat and get angry at his neighbors. Every day he quarrels at least once, and often two or three times, with each of his four or five neighbors. Each skirmish lasts anywhere from three to fifteen minutes. Richard D. Estes, who has extensively studied the behavior of the gnu, describes such a skirmish as a challenge ritual.[2]

The ritual begins when a bull crosses into his neighbor's territory, pretending that he has accidentally strayed over the border while he was busy grazing. Then the two opponents perform the following movements: running around each other in circles, smelling each other, drawing back their lips in a sort of aggressive "smile," urinating. Then both gnus kneel down, place their heads together, and push. They tear up the grass with their horns, shuffle and stamp their hooves, stand stiffly in the pose gnus assume when they are alarmed, buck like rodeo stallions, toss their hind legs in the air, and slowly calm down again. Eventually the intruder withdraws to his own territory, cropping grass in a leisurely manner to save face.

The gnu which is defending his territory is always the victor in the skirmish. That is, he always succeeds in driving away the intruder without engaging in active combat.

I mentioned that in the group courtship ritual of the northern mallard, the drakes must perform all their dance steps in a certain fixed order. However, among gnus, every ritual challenge is different. Each ritual begins and ends the same way, but in between, both gnus are left to their own devices. They may perform all or only a few of the aforementioned movements in whatever order, and with whatever emotional emphasis, they choose. Thus unlike the courtship behavior of northern mallards, the challenge behavior of gnus varies from one individual to another.

People often tell me, "Everything you have to say about the instinctual behavior of bees, mice, and ducks is very interesting, but it does not help us to interpret human behavior. Unlike animals, we are not controlled by our instincts."

Before replying to this remark, let me ask whether the gnu has also freed itself from the rigid corset of instinct. Why is the behavior of an individual gnu so much more flexible than that of a northern mallard? Can a gnu control its own behavior?

Gnus cannot deliberately control their behavior. During the challenge ritual, the two bulls vacillate between fear and aggression. Each movement a gnu performs expresses a momentary increase in fear or anger, in the desire to threaten or the desire to appease his rival. Many factors determine which emotion temporarily has the ascendancy. Among these factors are the gnu's own temperament, the temperament of his rival, his persistence, and how the two gnus have reacted to each other in the past.

Thus during the challenge ritual, the two adversaries do not in any sense control their instincts, but rather engage in an emotional duel. As they react to each other's behavior, they are tossed back and forth between feelings of fear and aggression.

Konrad Lorenz and his school explained instinct in terms of simple examples from the life of insects, fish, and birds. As a result, many people assume that instinctual behavior always follows very rigid patterns. This is not the case. Like gnus, human beings are torn between conflicting emotions. And like gnus, they may appear to be consciously controlling their behavior. In reality, they are always obeying their instincts.

Apparently, to maintain its emotional equilibrium a bull gnu must engage in constant quarrels with his neighbors. The challenge ritual never achieves any practical results. The two bulls do not actually fight, so neither of them is wounded or killed. Moreover, the intruder always retreats to his own territory, never making any serious effort to conquer that of his neighbor.

A bull gnu never attempts to establish a territory somewhere where he could be alone and live in peace. Instead, he seeks the proximity of his own kind, so that every day he will be able to quarrel with several other bulls.

Depending on the size of the gnu population, two bull gnus may be separated by an interval of from thirty to more than eight hundred yards. One might assume that bulls which do not live in close proximity quarrel less often than their more crowded peers. Such is not the case. Bulls that live eight hundred yards apart engage in ritual combat just as frequently as those that are stationed only sixty yards apart. However, it is true that bulls that are closer than sixty yards apart do quarrel somewhat more often.

Bull gnus use the challenge ritual to work off their pent-up aggression. If they do not release their aggression in this harmless way, they will release it on the females, driving them away and destroying their own chances to mate. Thus to maintain his emotional health, a bull needs the presence of an enemy as a whipping-boy. The bond between two enemies who need each other as whipping-boys is called a social attraction-rejection bond. No doubt readers will be able to think of people they know who need a whipping-boy in order to function.

How do young gnus succeed in wresting territory from such an aggressive, firmly rooted "establishment"? The attempt to win territory by brute force would be extremely dangerous. Moreover, even if such an attempt succeeded, four or five neighboring bulls would question the young bull's right to keep his new-won territory. Moreover, a young bachelor cannot establish territory where there are no neighbors, for all bulls need the proximity of their peers.

How does a bachelor gnu solve his dilemma? First he looks for a thinly settled and poorly defended area in the network of territories. When attacked by the landowners, he promptly retreats. However, he returns again and again, until finally the older bulls become accustomed to his presence and do not drive him away with such ferocity. The young male, growing more confident, begins to behave as if he already possessed a territory. As soon as a neighbor approaches, he withdraws, then

During combat, a bull gnu sinks to his knees to prevent the other bull from driving his horns into his vulnerable underbelly.

approaches again from the other side. He behaves like a sales-
man who, having the front door shut in his face, appears at the
back door smiling and unabashed.

The young bull becomes a serious threat to the territorial
rights of the older bulls when his self-confidence, and with it
his aggressiveness, grows to the point that he is willing to
openly defy them. Then the youngster engages in an endless
series of challenge rituals.

The mere fact that a newcomer is prepared to claim and
defend his rights exempts him from the need to fight in earnest.
Henceforth he can deal with the other bulls through the medium of the challenge ritual. Before he seizes a territory, the
young male's lack of aggressiveness prevents him from fighting
the other bulls. After he seizes a territory, his new-won aggressiveness and self-confidence cause him to behave in the manner
of the older bulls, which hardly ever fight in earnest. Only if the
newcomer lacks sufficient patience to allow the older bulls to
get used to him will he have to engage in a genuine battle.

In a serious fight, each bull tries to stab the other bull's
underbelly with his sharp horns. However, to prevent his rival
from doing this, before the battle each bull falls to his knees.
Their foreheads pressed to the ground, the bulls lock horns and
twist and shove until they almost dislocate each other's necks.
If one bull stabs the other in the neck with his horns, the blow
is usually harmless, for the necks of gnus are covered with a
thick horny skin. The two bulls may continue to fight for twenty
minutes. When one of the two is so exhausted that he can no
longer react swiftly enough to prevent the risk of a fatal wound,
he gives up and runs away.

With a few unhappy exceptions, the defeated bull does not
lose his life but only his property and his self-esteem. He now
has no choice but to remain a bachelor or to undertake the
laborious task of insinuating himself into a new territory.

The talent for acquiring territory varies from bull to bull.
Some bulls can acquire it in a few hours; others take weeks; still
others fail altogether.

Outside the Ngorongoro Crater, brindled gnus do not own
territory. Instead they are nomads. On the Serengeti Plain,

mating season falls in May and June, at the time of the great gnu migrations. In these months the bulls establish temporary mating territories.

Like a fleet of enemy ships, a herd of bulls will bear down on a train of migrating females and bar their path. Within fifteen minutes, the bulls divide the cows into groups of around sixteen cows to each bull. Then the bulls briefly stake out a territory and mate with the cows. After two hours or so, both bulls and cows will be swept into the stream of thousands of migrating gnus.

In many species of antelope and gazelle, only landowners are allowed to mate. The males of different species own varying quantities of land. The Defassa waterbuck has a territory of between twenty-five and five hundred acres; the male impala owns from fifty to two hundred and twenty-five acres; and the male Uganda kob, a species of antelope, is content with little more than one hundred square yards.

The amount of territory a male owns determines how he treats the females of his species. The male Defassa waterbuck and the male impala both own so much land that they can hardly see their neighbors. Both animals treat the females roughly and drive them back into the center of the territory if they try to run away. To keep the enslaved females from escaping, the males must guard them like sheepdogs. In effect, they keep a kind of harem.

Male Uganda kobs, on the other hand, have so little land that sometimes two males are as close as ten, twenty, or thirty yards apart. With so many other attractive males around, no male kob can keep a harem or treat the females like slaves. He has no choice but to be amiable. Thus before mating, he demonstrates no threatening or overtly sexual behavior which might frighten the females.

But whether they are cavaliers or grumps, male Uganda kobs, Defassa waterbucks, impalas, and gnus are all incapable of establishing a marriage. Territorial behavior makes it unnecessary for male and female to develop a close bond. After mating, males and females immediately go their separate ways. The females rear the young alone.

# CHAPTER 14

# Beauty Disqualifies Males from Marriage

*Arena Behavior*

Human beings always assume that our species represents the crowning achievement of the creation. Thus because monogamy constitutes the ideal form of human marriage, we assume that monogamy must be the most highly developed form of marriage in the animal kingdom and therefore must have been the last to evolve. In some species it is true that monogamy appeared long after other forms of sexual union. However, some initially monogamous species later developed nonmonogamous trends. In these species, the female will have nothing to do with the male after the two have mated.

Birds exhibit a variety of evolutionary trends. In general, bird species have tended to develop away from promiscuity and toward permanent, monogamous sexual relations. For example, jackdaws, ravens, and greylag geese are now monogamous species that mate for life. The marital problems of these birds closely resemble those of human beings.

However, some birds have evolved in the opposite direction: Initially monogamous, they later ceased to engage in pair formation. This group is composed of species in which the males adorn themselves with bright feathers to please the females. The females leave the males immediately after mating, for the male's colorful plumage would attract predators to the nest and endanger the young birds.

The form of sexual relations in a given species profoundly affects the future evolution of that species. In some bird species, the females select the most brilliantly colored males as mates. Thus the male birds have gradually developed increasingly striking and beautiful plumage. However, the more beautiful the male is, the faster the female drives him away after they have mated. Once the proud suitor has fulfilled his biological function, the female does not care if he is eaten by a predator—as long as it does not happen near her nest. Thus birds of this kind do not "marry." In terms of the evolution of behavior, they must be considered "backward."

Clearly, if the human male had been the more beautiful of the sexes, the marital relations of human beings would be very different from what they are today.

I will now discuss one phase in the evolution of certain monogamous bird species into species which do not marry. This phase is represented by the black grouse.

It is an April morning shortly before dawn on Germany's Lüneburg Heath. A game-keeper named Heinrich Hesse has constructed a blind of pine branches between the clumps of juniper and furze.[1] Since time immemorial the male black grouse have been coming to this spot the year around—especially in April and May—to display and impress each other with their beauty. As yet there are no birds in sight.

Suddenly the first male grouse lands about fifty yards from the blind. He looks around carefully. Then he begins blowing, hissing, and cooing. He has hardly begun his performance when a second grouse lands right next to the blind. At once the first grouse flies over to the second, as if he did not enjoy performing without an audience. Soon five grouse have gathered in the area so that they can be unfriendly to each other. Like the males of many species, they need the presence of rival males in order to function.

His head thrust forward aggressively, his tail-feathers spread, and the white feathers under his tail forming a glowing circle, each grouse begins to run around cooing loudly. His bright red combs[2] swell and poke out through his feathers over his eyes. The air sac at his throat swells, acting as an amplifier. Thus each time the bird springs in the air, his call, "Chooee-ee," resounds two miles across the heath. At the same time the grouse beats his wings against his legs like a man slapping his thighs in a Bavarian peasant dance. (The Bavarians based one of their dances on the movements of the black grouse.)

All the grouse pretend to attack each other, drumming on the ground with their feet and sounding battle-cries: "Rutturu-ruttu-ruiki-urr-urr-urr-rrutturu-ruttu-ruiki!" However, their behavior is all sound and fury. They have no intention of fighting.

The moment the sun comes up, the birds stop puffing like steam engines and pause in their mock duels for a moment of silence. Hunters call this moment the "morning prayer." Then the grouse resume their theatrical display. Eventually a plain-looking brown female may appear on the scene. As a rule, few females come to the dancing court, for they need to mate only once. The males, on the other hand, come here to display their beauty every day.

Seeing the female, the males duel more ferociously than ever. Finally the female selects the most striking and colorful male of the lot and crouches down in front of him, inviting him to mate with her. When the sexual act is completed, so is the "marriage." The brown female scurries away and builds a nest somewhere in the bushes. Several days later she lays around eight eggs, hatches them, and raises the young birds alone. Once the beautiful male has served his purpose, the female must carry on alone, for the male's beauty would attract predators and endanger her young.

Meanwhile, the "most beautiful male in all the land" continues to strut around every day on the dance floor. Naturally, all the females invariably choose him as their mate. The other, less beautiful males lead the dreary life of wallflowers.

When only the most beautiful males reproduce, a species rapidly evolves more and more beautiful, and thus unmarriageable, males. Eventually the males are no longer able to fly, keep the eggs warm, or perform any useful function. They are good for nothing but to strut around in front of the females.

The black grouse has many relatives belonging to the grouse subfamily. Members of this subfamily gradually evolved males with increasingly beautiful plumage. The hazel grouse represents the first stage in the evolution of the purely decorative male. In this species, the cock and the hen both have plain coloring that serves as camouflage. The two sexes closely resemble each other. However, in autumn of the first year of his life, the young cock performs a courtship ritual that makes him look like a rather clumsy apprentice of a male black grouse. Crying "Tzi-tsitseri-tsitsi-tswi," he spreads his tail-feathers, waves his tail up and down, spreads his wings with the tips drooping toward the ground, and makes the bright red combs over his eyes swell so that they poke out through his feathers.

After his courtship display, the male hazel grouse looks as plain and behaves as unassumingly as ever. Thus his presence poses no threat to his mate, his nest, and his children. After a six-month "engagement," the cock and the hen mate the following spring. They spend the six months of their engagement together. After mating, they raise their seven to ten offspring together and remain faithful to each other until one of them dies.

The ptarmigan represents the second phase in the development of the purely decorative male. The cock's plumage is predominantly plain, but he possesses a row of large black and white feathers. Moreover, unlike the male hazel grouse, the male ptarmigan does not privately court a single female. Instead, in spring the cocks gather in groups and compete for the attention of the females.

When male birds gather in groups at appointed places to display themselves to females like performers in a circus ring, their behavior is called arena behavior.

Each male ptarmigan selects a rock, a bush, or some other prominent point in the landscape to use as a pedestal. Unlike the male black grouse, which engages in mock combat, the ptarmigan fights in earnest to keep other males away from his perch. At this stage of evolution, the behavior of the cocks has not yet been completely ritualized. Thus they still engage in genuine combat.

In his courtship display, the male ptarmigan whirs thirty feet or so into the air and makes a wide circle before returning like a boomerang to his own perch. He may travel as much as one hundred and twenty-five feet around the perch. As he flies through the air he makes as much noise as possible, and lands with a shrieking "Arrr-arrr." The males perch some distance apart. Thus each male must fly around and make noise to ensure that the females see him and know that he is looking for a mate.

When a female selects him as her mate, the male ptarmigan follows her from a distance. His plumage is too colorful to serve as ideal camouflage. On the other hand, it is not so colorful that it is dangerous for the female to be seen near him. Thus at first the cock and the hen conduct a sort of long-distance marriage. Then the hen builds the nest alone, and between twenty-one and twenty-four days later, she lays her six to ten eggs and begins to sit on them. Meanwhile, the cock stands guard some distance away, ready to defend his family from small enemies and to divert the attention of large predators from the nest.

As soon as the young birds hatch, the cock withdraws even farther from his family while he grows his autumn plumage. He may disappear altogether for a week or two. His new coat of feathers serves as excellent camouflage. When he is as drab-looking as his wife, he returns to her and his almost fully grown children. Now he can stay with them without endangering their lives.

Clearly the male ptarmigan is not consciously aware that when his feathers change color, it is safe for him to return to the nest. Presumably, instinctual fluctuations in his feeling of attraction to his family determine how close he comes to the nest. Through natural selection, these emotional fluctuations have become linked to changes in the color of his plumage.

In winter, all ptarmigans grow a third coat of feathers—a white coat that makes them blend into the snow.

The capercaillie, the black grouse, and the ruff represent the

The male sage grouse is unique among the males of the world in that during his courtship display, two "breasts" appear on his chest.

next stages in the evolution of the beautiful but useless male. However, by no means do the males of these species represent the ultimate in masculine beauty. The sage grouse of the North American prairies puts on a far more spectacular show. Male sage grouse meet in an arena over eight hundred yards long and more than two hundred yards wide—an area comparable in size to the Circus Maximus of ancient Rome. Around four hundred cocks assemble in this giant arena.

Each male stakes out a private territory in the arena, spreading his tail feathers, holding his wings in front of him like a shield, and inflating the enormous white air sac at his throat. His white feathers tremble so that they make an audible sound. Then two orange-colored swellings the size of oranges poke out through his breast feathers. Among sage grouse, not the female but the male has "breasts."

Suddenly the male grouse utters a piercing call that resounds more than three hundred yards across the prairie, and his inflated body collapses like a pricked balloon. Then the game begins all over again.

In the center of the arena stand four "master cocks," the cocks that the females most frequently choose as mates. These four cocks mate with seventy-four per cent of the hens. When they are exhausted, a half dozen "vice-masters" take over thirteen per cent of the remaining females. Another thirteen per cent of the hens choose lower-ranking cocks. Around three hundred and fifty of the four hundred cocks in the arena will never be able to mate with any hen.

Obviously, these inflated male bundles of feathers do not make suitable marriage partners. Once they have mated with a hen, they ignore her and turn their attention to the next hen. Meanwhile, the females depart to build their nests, lay their eggs, and raise their young.

The better a male bird gets along with females, the better he can get along without them. Male sage grouse make even less suitable mates than male black grouse.

Paradoxical as it may sound, the female birds themselves are responsible for the creation of these useless and decorative males. The females are exclusively interested in the beauty of the male's posture and feathers and are unconcerned with his strength, ability to fly, or survival skills. Nor do they place any value on mutual sympathy or emotional compatibility. The females' taste for gaudy and ornate males accelerates the tempo of evolution, rapidly creating males that are unable to fly, defend their families, keep eggs warm, or do anything except be attractive to females.

Thus in the course of biological evolution, species do not always become better adapted to their environment. As a rule, natural enemies, competitors, climatic conditions, and other environmental factors govern natural selection. In rare cases, factors like a female ideal of male beauty intervene. When this happens, a species can develop traits that may prove disadvantageous in the struggle for survival.

As human beings, it may please us to know that aesthetic as well as utilitarian factors can govern evolution. On the other hand, if a species develops too many disadvantageous traits, it may become extinct. It is important for us to remember this biological law, for like the sage grouse, we do not always govern our societies with a view to ensuring our survival as a species. Instead we are dominated by social, ideological, and political ideals and obsessions that may in the end endanger our survival.

The enormous fangs of the saber-toothed tiger were responsible for its extinction. These fangs were almost five inches long. Equipped with such long fangs, the tigers' jaws no longer served their basic purpose: killing and eating. Probably the males developed these extravagant decorations because they pleased the females.

At the end of the last Ice Age, the antlers of the male European giant deer measured almost four yards from tip to tip. At that time, the landscape of Europe resembled the arctic tundra. These gigantic antlers served no purpose except as insignia of rank. Later, when vast forests covered the land again, the antlers that so impressed the female deer made it impossible for the stags to survive. As a result, the species became extinct.

Moreover, it may have been this same kind of sexual selection that caused the dinosaurs to develop such huge unwieldy bodies, inevitably bringing about their extinction.

When a species develops impractical or disadvantageous traits, must it necessarily become extinct, or is there a way for it to escape this dead end? In other words, do beautiful males doom a species to extinction?

The most beautiful of all birds belong to the family of peacocks, great arguses, superb lyrebirds, and birds of paradise. For good reason, these paragons inhabit regions where there are relatively few predators. Birds of paradise are to be found only in the tropical rain forests of New Guinea, northern Australia, and some of the surrounding islands.

Describing these loveliest of all birds, E. Thomas Gilliard states that the males occupy courtship territories and perform fantastic tricks for the females, spreading their tails, mantles, fans, trains, wattles, and colorful crests.[3] Often they open their beaks wide to display the glowing green or opalescent linings. During display, the birds stretch as tall as they can, dance forward and backward, swing themselves in circles around the branches where they are perching, sway from side to side, and hang upside down. Some slowly unfold their feathers like the petals of a flower; others open and close their fans, feathery beards or bright shields. The King of Saxony's bird of paradise displays his giant flag of feathers; Arfak six-wired parotias dance in the underbrush; and other birds of paradise perform gymnastics around the trunks of spindly saplings. The true birds of paradise wave their feathers like fluttering trains.

There are around forty species of birds of paradise. These species exhibit the same evolutionary trend as the grouse subfamily: the trend toward increasing beauty. Some species, including the multicrested bird of paradise, the green-breasted manucode, and the trumpet bird, have not evolved colorful males. The males of all these species, which are probably descended from crows, are quite plain looking, and the males and females closely resemble each other. As a result, after mating the cocks and hens remain together and live in permanent monogamy.

However, like the colorful males of some grouse species, the more brilliantly colored male birds of paradise, not content with being beautiful, all want to be more beautiful than the other males. They do not try to outshine each other in strength, but only in beauty. To challenge their rivals, they meet in an arena to conduct a communal courtship.

Among the plainer birds which court in an arena are the

If a peacock wants to mate with one of the four or five hens in his harem, he does not openly pursue her. Instead he goes to the traditional courtship site in his territory and, even if no female is present, spreads his magnificent tail. At one time scientists believed that the spreading of the tail was a ritual signal that the peacock wanted to mate. However, in reality the signal means that food is present. If a hen responds to the call, she does so only because she wants food. Far from being dazzled by her mate's beauty, she is disappointed when she finds that he was shamming and that there really is no food. Despite her disappointment, the male does sometimes succeed in mating with her. Thus a peacock courtship is based on a misunderstanding: He is thinking about love, but she is only interested in eating.

# CAPTIONS FOR THE COLOR INSERT THAT FOLLOWS

The male northern gannet beats his mate several times daily. These seabirds are approximately the size of geese. Every day, a northern gannet couple reenact the history of sexual relations, the evolution from aggressive behavior to sympathy and mutual tenderness. Each day when one of the mates returns to the nest after catching fish, the male attacks the female. Gradually the thrusts of his beak grow gentler until he is only pretending to hurt her. In the end, the two mates caress each other by running their beaks through each other's feathers.

In his courtship dance the male crane tries to impress the female by leaping high into the air. Native tribesmen in many regions of the world have imitated the dance of the crane to impress their womenfolk. The picture shows the crowned crane, whose dance has been imitated by various black tribes in East and South Africa. Canadian Eskimos modelled a dance on the courtship dance of the American whooping crane. The Japanese Ainu imitated the Manchurian crane, and the Australian aborigines the Australian crane. The courtship dance of the crane is unusual in that the same gesture—the leap into the air—impresses the females of many different species.

Greater spotted woodpecker mates do not like each other. Except when mating, they remain some distance apart. Nevertheless, both male and female are devoted parents.

The mass courtship of flamingoes is among the most impressive sights in the animal kingdom. Hundreds or thousands of the graceful long-legged birds perform a stately dance, moving their necks and wings in ritual gestures and uttering rhythmic groans that come from deep in their throats. The sexually mature flamingoes form a separate group within the flock and dance together. Within this group the males and females select their mates for the current breeding season.

A pheasant marriage is a bizarre phenomenon. At the beginning of April, two or three hens join the colorful cock (*on the right in the picture*). Then, accompanied by his harem, the cock goes walking through his territory, always following the same route. Periodically he invites one of the females to mate with him by pausing in front of a tasty morsel of food and offering it to the chosen hen. As he offers the food, he seductively calls, "kutye, kutye." If the hen accepts the food, the two birds mate. Strangely enough, the cock does not acually court the hen until after they have mated. Then he circles around her with his head lowered almost to the ground, spreads his wings, and displays his brilliant plumage. Around fourteen days later, when the hen lays her first egg, the harem disbands. The male's bright coloring might attract enemies to the nest and endanger the chicks. Thus once the eggs are laid, the male is banned from the nest.

Female gnus select the males they wish to mate with. The picture shows a herd of females with their young. In mating season these herds enter the region where around two thousand bulls have staked out their territories. Then the cows form smaller herds of around twenty cows each. Each herd enters the territory of the bull that the majority of cows seem to prefer. Thus the cows choose their mates collectively. If a bull behaves badly, the cows leave his territory and move on to that of his neighbor.

Male gorillas may weigh as much as six hundred and fifteen pounds and when erect may reach a height of seven and a half feet. A gorilla has the muscular strength of four or five men. Because of their size and strength, it was once assumed that gorillas had insatiable sexual appetites. In reality the sexual drive of gorillas is relatively weak. The females care for their young for between three and a half and four and a half years. Thus they mate only once every few years. Gorillas live in groups. As a rule each group contains a silver-backed old chieftain, two adult males, six or seven females, three "teen-agers," and four or five babies. Gorillas practice free love. When a female is in heat, the leader never attempts to prevent her from mating with any male she chooses. Unlike male chimpanzees, male gorillas do not rape the females.

Bullfinches are monogamous and mate for life. Before mating, they become engaged twice. The first time, they are engaged to a brother or sister. This first engagement trains them how to behave toward their mates. The second engagement is a test of compatibility. If two bullfinches are not compatible, they break the engagement, and each becomes engaged to another bird. A mysterious instinct prevents brothers and sisters from mating.

twelve-wired birds of paradise. Actually, in comparison with most bird species they are far from plain. Their bodies are like lemon-yellow muffs. They have small brown tails and their heads are small and black, but their necks are ringed with a brilliant turguoise and emerald-green frill. At dawn this "Lord High Steward" perches in the summit of a tall tree in the New Guinea jungle and uttters a loud cry that awakens his neighbors. Soon the other males follow his example and perch in the adjacent trees.

At times the male birds may be as much as several hundred yards apart. Probably they cannot even see each other. However, they *can* hear each other's cries, and the cries of their rivals stimulate them to perform their courtship dance. Far apart as they are, their behavior typifies that of males displaying in an arena.

The greater bird of paradise is a master of the art of the courtship dance performed in a smaller arena. The males are bright, fluffy balls of golden feathers wrapped up in snowy-white, misty tulle. As soon as one of the adult males, trembling with excitement, puffs out his feathers and begins to parade from branch to branch, a number of young males flock around him. At this age, they are not yet as magnificent as he is. All the same, the eager apprentices attempt to imitate his graceful movements.

The greater bird of paradise performs before dawn. It seems paradoxical that the males adorn themselves with bright plumage to please the females and then hide their light under a bushel, dancing before the sun rises to shine on their feathers or in the dark shade of a tree in a jungle thicket. Perhaps they fear their own beauty and the predators it might attract. However, since they dare to dance only in dim light, their plumage must be all the brighter in order to impress the females. Thus they are caught in a vicious circle.

Another master of the techniques of arena courtship is the red bird of paradise, an enchanting cascade of red, yellow, green, brown, and black feathers. As many as forty of these beautiful creatures may simultaneously perform a ballet in a jungle tree. The dancing birds pursue each other through the branches, then abruptly pause and posture, draping their fiery veils around their bodies.

The small drab females may watch the performance for a long time without showing a spark of interest. They look as if were fed up with the tedious display of feathers. A male has to turn in a truly spectacular performance before a female will walk up to him and select him as her mate.

Some New Guinea tribesmen have been so impressed by the mass courtship of birds of paradise that they have imitated the courtship dance in their tribal ritual. The men adorn themselves with bird of paradise feathers and dance together in an "arena." The man wearing the most beautiful feathers is the most attractive to the young women of the tribe, even if the man himself is not as strong and handsome as his rivals.

For decades the birds of paradise have held their courtship dances at the same locations. Various Papuan families now own the desirable strips of land where the birds display. Thus the sons of the most powerful, or land-owning, families have access to the most and the finest feathers, and the "best-dressed" men in the arena are in fact the wealthiest and most influential men in the tribe. In other words, the women of New Guinea are very much like women everywhere in that they are attracted to wealth and power.

Naturally, the most colorful birds of paradise are the ones most frequently attacked by predators and human beings. So let us return to our original question: What happens to a species that the aesthetic whims of females cause to develop disadvantageous traits? In short, are the birds of paradise doomed to extinction?

The answer is no. These birds are not becoming extinct. E. Thomas Gilliard [4] believes that birds related to birds of paradise are the ancestors of the legendary bower birds. That is, birds of paradise have evolved in a new direction.

Some bird of paradise species are intermediate between birds of paradise and bower birds. That is, they represent transitional species. They are the magnificent bird of paradise and the Waigeu bird of paradise, species in which the males whirl like small helicopters around the trunks of young saplings. A step closer to the bower bird is the Arfak six-wired parotia, which waits until a female approaches and then flies to the ground to perform his ballet. Spreading his black feathers to form around his body a circular fringe resembling a ballerina's tutu, he hops two steps forward and two back and then swiftly revolves in a circle. The males of all three of these bird of paradise species spend some time cleaning up their dance floor before the performance, clearing away all the plants that might block the female's view. They also pluck all the leaves from the branches overhead. Often their "stage" has a diameter of from five to seven yards and may be over thirty feet high. Thus the male dances in full view of the female and in a "spotlight" of sunshine that streams down through the leafless branches and illuminates his bright feathers. This special lighting technique makes it unnecessary for the males of these species to wear such bright colors as those birds of paradise that dance only in dim areas of the jungle.

Thus a change in behavior patterns can halt the trend to develop more brightly colored plumage.

Other birds species have developed behavior patterns that helped to halt the evolution of more brilliant plumage. In fact, these patterns have actually caused the birds to develop plainer coloring.

The crucial behavior pattern is the plucking of leaves. Obviously, to help his mate build a nest, a male bird must be able to tear up grass and leaves and weave them together. Very beautiful males lose the ability to build a nest; or rather, the nest-building instinct becomes dormant.

The drive to pluck leaves has been reactivated in the male magnificent bird of paradise and the Arfak six-wired parotia. These birds expend enormous energy to clear of leaves a space as large as five to seven yards in diameter and as much as thirty feet in height. In comparison, it is a small task to build a nest.

The drive to pluck leaves is a variation of the nest-building instinct. In the bower birds, we see the effects that such a drive can have on the evolution of a species.

After the magnificent bird of paradise and the Arfak six-wired parotia, the courtyard bower birds illustrate the next

The birds of paradise evolved increasingly ornate plumage: (1) the crowlike trumpet bird, (2) the twelve-wired bird of paradise, (3) the red bird of paradise. Number 4, an Arfak six-wired parotia, represents a transitional form between the birds of paradise and the bower birds. Bower birds grad-

ually evolved increasingly elaborate bowers, or courtship areas. Since the bower attracts the female, the male himself can afford to be relatively plain. Number 5 is the yellow-brown golden bower bird, and Number 6 is the plain-looking brown spotted bower bird.

stage in the development of more plainly colored males. Among the courtyard bower birds is the Archbold bower bird.

The Archbold bower bird prepares a courtship area some two to three yards in diameter that he lines with ferns. Every day he removes the withered foliage and adds it to the wall surrounding the courtyard. Most significantly, on top of the wall he piles heaps of "treasure"—a bunch of brightly colored berries, the iridescent armor of insects, snail shells, lumps of resin, charred wood, and orchids or other flowers. In this species, the beauty of objects external to the male begins to serve as a substitute for the beauty of the male's plumage.

The magnificent rifle birds, which belong to the birds of paradise, also exhibit a fondness for decoration. The females decorate their nests with cast-off snakeskins. However, it is not clear whether they actually like the way the skins look or simply wish to frighten other snakes away from their nests.

There are eighteen different species of bower birds, one more ingenious than the next. Tower-building species are not content merely to line their courtyards with leaves or ferns. They actually build towers. The golden bower bird of northern Australia piles heaps of branches around the trunks of young trees and then weaves them together to form a tower almost ten feet tall. For a human being, this would be like building a tower two hundred and sixty feet tall.

Male golden bower birds try to build taller and finer towers than their neighbors the way Manhattan business firms try to build taller skyscrapers than their competitors. The males are still competing in an arena, but now they compete with buildings instead of feathers.

Some gardener bower birds actually build several towers side by side and connect them with walls of twigs which they decorate with beautiful objects. Other species build houses shaped like the tepees of some American Indian tribes. The houses contain large rooms, and there are protective barriers in front of the doors.

These architectural wonders do not constitute nests where the females will lay their eggs and raise their young. Instead they are love-nests, used only during courtship and mating. If

the female birds had a hand in the building, they would undoubtedly produce something more practical.

The purpose of the beautiful buildings and their decorations is to show off the male's physical beauty to best advantage.
However, the buildings are so much more magnificent than the
architects that the males' appearance is relatively unimportant.
As a result, the males themselves can afford to have plainer
coloring.

However, plain though male bower birds may be compared
to male birds of paradise, they do not represent suitable marriage partners. Immediately after the two birds have mated, the
female flies away, builds a plain, ordinary nest like any other
bird, and raises her young alone.

The yellow-breasted bower bird rams several thousand
little sticks and twigs in the ground, weaves rushes through
them to form a solid wall, and lines the enclosure with grass.

The male satin bower bird actually paints the inside of his
bower. He takes a piece of bark or a dry leaf and frays one edge.
Then he holds the bark or leaf in his beak and uses it as a paint-
brush. Depending on the species, he may use blue or green
paint. He makes the paint by mixing the juice of berries with
saliva. When the paint is ready in his beak, he holds the "brush"
with the frayed edge downward and lets the paint run onto the
brush. Then he rubs the frayed edge up and down along the
walls.

The female bower bird selects her mate not for his size,
strength, or character, nor for the beauty of his plumage, but
rather for the skill, industry, and creativity with which he builds
their love-nest.

How does skill in building bowers contribute to the survival
of bower birds? Would it not be ironic if there existed a species of bird which, unlike most human beings, revered artistic
skill above other, more material values?

Regrettably, physical strength plays a major role in the
artistic achievements of bower birds. If a builder is skilled and
industrious but physically weak, he will lose out in the struggle
to attract females. Like all neighbors, the neighbors of bower
birds tend to be covetous and spiteful. If a strong bower bird

sees that one of his weaker neighbors has built a taller and more beautiful building than his own, he will come over and tear it down.

The more complex the architecture and the richer the decoration, the plainer the coloring of a male bower bird will be, and the better camouflage his plumage affords him.

Male bower birds execute complex courtship dances involving certain traditional steps, leaps, and movements of the wings. However, during their dance the males do not display red throat-feathers or crests, but carry large red berries in their beaks. This behavior is a variant of courtship feeding. Among bower birds, complex behavior patterns have replaced physical appearance and other types of sexual signals as a method of inducing the females to mate. Thus the evolution of the bower birds has saved the birds of paradise from extinction.

The European giant stag became extinct, but its descendants, the modern deer, did not. The modern deer and the bower bird show that the development of disadvantageous traits does not necessarily doom a species to extinction. Like the male bird of paradise, the male bower bird has continued to develop increasingly magnificent displays for the purpose of impressing females. However, in the bower bird the trend toward increasingly elaborate display is expressed through the creation of objects separate from the male himself.

Thousands of years from now, the descendants of the present-day bower birds may once again marry and live permanently with their mates. The drab plumage of the male bower bird poses no threat to the nest. Thus there is no reason why, at some time in the future, he could not share in raising the young.

The bower bird, which is evolving toward monogamy, is a later form than the birds of paradise. This fact shows that in birds, monogamy is not as "backward" a form of sexual relations as it may have first appeared.

# CHAPTER 15

# Courtship at Court

*The Turkey System*

I have discussed various techniques by which male animals enhance their beauty or erect beautiful buildings in order to be attractive to females. In 1970 zoologists discovered that some male birds keep a court of other males to magnify their personal charms.[1] A turkey society is a veritable dictatorship.

The scene was the desert in Welder Park, a wildlife preservation park near the town of Corpus Christi in southern Texas. Like four fully rigged sailing ships, the turkey "General Lee" came marching along with his three brothers, John, Jim, and Jack. Their tail-feathers spread, their wings spread and touching the ground, their feathers fluffed out, the red wattles at their throats bulging, and gobbling loudly, they trotted through the sere landscape toward a group of fifty-two hen turkeys.

From the left and right, two other groups of three males each were closing in on the hens. Meanwhile, at the rear several two-man groups and various solitary males were doing all

**245**

they could to impress the hens with their grandeur. Then, like a fleet of ships preparing to engage the enemy, General Lee and his three brothers approached the nearest group of three males. When the "Lee brothers," still in military formation, came within twelve feet of them, the other three turkeys lowered the flag without a fight. Docilely folding their feathers, they crept away looking small and crestfallen.

The mere approach of General Lee and his confederates soon compelled all the other turkey units to capitulate. The defeated turkeys promptly retreated to the sidelines to watch the victors approach the hens.

The General and his three brothers paraded like majestic Indian chieftains among the flock of hens. Finally the General saw a hen he liked and began to court her. Meanwhile his three brothers surrounded the hen, spread their wings and feathers, and gobbled just as enthusiastically as their domineering big brother. The younger brothers knew perfectly well that they would not be allowed to mate with the hen. They were not courting her for themselves but merely enhancing the prestige of their brother.

The General and the hen had just begun the ritual of mating, which lasts for from four to five minutes, when a raccoon darted out of the bushes. The other male turkeys, doomed to stand around the hens and watch, saw the danger, and ranged themselves in battle array in front of the hens and the General. Beating their wings, they gobbled ferociously at their enemy and appeared determined to attack him with their beaks and the sharp spurs on their legs. Soon they drove away the raccoon. Meanwhile, the General calmly continued to take his pleasure.

During the next four weeks, much the same scene was repeated over and over. A total of thirty-one males all gathered around the fifty-two females. However, only General Lee ever had a chance to mate with the hens. He mated with them one after another. Whenever he was present, all the other males stood around like extras in a movie.

Occasionally a turkey cock on the far side of the flock of hens would make a surreptitious attempt to mount one of them. At once the squad of four brothers would bear down on the unruly fellow to put him in his place. The courtship ceremony preceding copulation always takes four minutes. Thus the General, aided by his staff, could always spare a few moments to quell a rebellion before rushing back to his bride. By the time he had finished teaching the rebel a lesson, the hen was emotionally prepared to mate with him. Thus quite unintentionally, renegade cocks helped to achieve the sexual synchronization of hens with the dominant male. Twenty-seven of the cocks served no other purpose than to indirectly help prepare the hens for mating, and to defend them against the attacks of enemies.

The General's three brothers were also doomed to sexual abstinence. Their task was to help their leader maintain his supremacy over all the other groups of male turkeys, and to help him court the hens. Thus the sexual behavior of male turkeys resembles the behavior of a motorcycle gang in which the lower-ranking members of the gang must help their leader to court his girl friend by running their motors and showing off the leader's power.

This wild turkey society in southern Texas consists of slaves and one all-powerful potentate who monopolizes the females. One male mates with fifty-two females while thirty other males stand helplessly by. None of the males ever seriously rebels. The dominant turkey does not even have to defeat his rivals in battle. He need only assume a dominant posture and *look* victorious.

How did General Lee first come to power, and why do the other cocks not dare to resist him? To answer these questions, G. Robert Watts and Allen W. Stokes studied the life of wild turkeys.[2]

Hen turkeys lay as many as fourteen eggs and sit on them until they hatch. By now readers will have sufficient insight into the marital psychology of animals to know that cock turkeys are incapable of helping the hen build the nest, hatch the eggs, or raise the young. That is, they are incapable of marriage. The females, of course, are quite capable of fulfilling marital obligations. In the animal kingdom, only males ever make unsuitable marriage partners.

At the beginning of April, the young turkeys hatch. In the

first few weeks, predators and the harsh climate reduce their
number by half. Soon groups of mothers band together with
their young. The young of different mothers tend to get into
fights. The group that wins is the one containing the largest
number of young cocks. A group of chicks including six females
and one male will be much weaker than a group of one female
and three males. At this age, males and females have equal
strength, but the males are much more aggressive. The mothers
never intervene in the quarrels of their young.

At the beginning of December, all the young cocks leave
their mothers and sisters and begin to live with their brothers.
Their attachment to their mothers and sisters disappears. The
attachment is replaced by a contempt for females, which in
later life periodically yields to the sexual drive. Henceforth
cocks feel close only to their own brothers. Two, three, four,
and sometimes as many as five brothers will be faithful to each
other for life. If there is only one surviving brother in a brood,
he will remain alone, for no outsider is ever accepted into a
turkey brotherhood. However, all the brotherhoods belong to a
larger group composed of young cocks.

Like many male animals that need other males as whipping-
boys, young male turkeys engage in innumerable battles. These
battles determine the social status each turkey will have for the
rest of his life. First brothers fight for supremacy, beating their

The picture shows a California sea lion proudly surveying the eight
females in his harem. All the cows are in awe of him, and he must court
each one for at least a day before she will consent to mate with him. When
he is worn out by the constant sexual demands of the cows, another, stronger
male will take his place. Thus a harem changes owners several times each
mating season.

Two of the cows in an elephant seal harem are fighting. The cows do
not like living in groups and would escape if the bull did not prevent them.
Depending on his strength, a bull may be able to keep as many as twelve
cows under control. Small wonder that the cows, crowded together in a small
space, are so frustrated that they quarrel among themselves.

wings, slashing at each other with their spurs, and jabbing at each other's heads and necks with their beaks. The fight is always fair: Two brothers never gang up against one.

The battles of cock turkeys resemble the battles of roosters in a chicken yard, except that turkeys use their wattles during combat. Many people may have wondered what purpose wattles serve, for they appear to do nothing more than get in the turkey's way when he is eating. Wattles are extremely leathery and tough. When they fight, two cocks grab hold of each other's wattles with their beaks and then twist them back and forth as hard as they can. Sometimes a battle continues for more than two hours. It does not end until one of the opponents is completely worn out. As a sign of capitulation, he lies down in front of the victor and stretches out his head and neck so that they lie flat on the ground. In this position, he could easily be killed if the other turkey chose to jab him in the throat. However, the victor will not kill his brother, for he will need him later to help him fight his battles and court the hens.

Brother fights brother until the hierarchy of power within the family is established. Then each brotherhood must fight the other brotherhoods. Few animals engage in such violent combat as turkey brotherhoods. Instead of fighting one to one in fair combat, two or three cocks will now attack one. Thus invariably the largest group, usually a foursome, defeats all the other groups.

Once ranking order has been established, a turkey society lives under a sort of *pax romana*. There is no need for the birds to fight again. Even when changes occur in the relative strength of the brotherhoods—i.e., when some of the brothers die—turkeys do not alter the original ranking order.

In the Texas desert, the mortality rate among turkeys is very high. Forty per cent of the birds die each year. Two brothers of General Lee had died when they were still chicks. Later that year, John and Jack died too. The General was now left alone with only one of his brothers. He was no longer more powerful than the three-man brotherhoods in the flock. Nevertheless, he retained his position until his death, when the remaining groups began to fight for supremacy.

The ruler of a turkey society is the least likely to die, for he is always among the females and has the most plentiful food supply. His underlings are in much greater danger, for they must stand some distance away from the females, guarding them against predators.

A turkey society is a sexual monopoly dominated by a male that is not necessarily the strongest or most beautiful. The society subsists through the labor of underlings whose self-sacrificial behavior recalls that of the denizens of insect states. Why did such a society develop?

Whenever food is scarce or there are so many young that a single parent cannot feed them alone, a species must develop monogamy in order to survive. The American quail exemplifies such a species. However, when food is abundant, a species is likely to develop polygamy, polyandry, or harems, and the males may practice arena behavior.

A second factor that helps to determine the mores of an animal society is the landscape. Animals living in a region where there are many hiding places will be safest if they live alone, in twos, or in very small groups. In this way they will have no trouble finding shelter or places to build their nests. Larger groups would simply attract the attention of predators. Examples of species that live in small groups are the capercaillie and the ruffed grouse.

The ruffed grouse conducts a "long-distance" marriage. Even though he is not brightly colored and thus poses no danger to the young birds, the male ruffed grouse keeps watch some distance from the nest. However, he is quite capable of caring for the young. If his mate dies, the male grouse will sit on the eggs, hatch them, and raise the young birds himself.

On grassy plains there is much less shelter than in the forest. Many birds congregate in the few areas where there is enough shelter for them to build nests and raise their young. Thus in these areas, most species live in large groups. For example, both prairie chickens and weaver birds live in colonies.

The Texas desert affords a plentiful food supply but little shelter. Thus conditions favor the development of large societies, like the turkey society, which practice polygamy. However, why do turkeys practice polygamy rather than the harem system?

The third factor which influences the mores of animal societies is the weather. In southern Texas it rains very seldom and at irregular intervals. However, plants and insects thrive for a brief period after a rainstorm. Turkeys must raise their young during this period, when food is abundant. That is, the hens must be prepared to mate and lay eggs as soon as it rains. Emotionally and physically, a hen must be synchronized with her partner as rapidly as possible. When a group of males like General Lee and his brothers all court a hen at the same time, they achieve this synchronization almost instantaneously.

Turkeys behave differently in different climates. Even minor changes in the environment produce major changes in the social order.

Another species of turkey lives in the forests of Georgia. Forests do not favor the development of large animal societies. Thus here the turkeys live in small groups. A cock has only one or two, or at most three hens in his harem, and the cocks and hens separate after mating. The cock is incapable of marriage.

Turkeys in Oklahoma have a third type of social order. Here they live on the prairie where the weather is not as dry as in Texas. Thus it is unnecessary for the cocks to mate with the hens as soon as a rainstorm begins.

The Oklahoma cocks attempt to establish the same sort of dictatorship as their relatives in southern Texas. But here the females do not always cooperate. In Oklahoma, it rains at fairly regular intervals. Thus the hens are prepared to mate only in a certain season. At other times of the year, they reject the advances of the turkey brotherhoods. When mating season arrives, a number of hens almost simultaneously signal their readiness to mate. At this point the cocks all become so excited that, ignoring the protests of the potentate, each cock tries to win a few hens for himself. The dictatorship is overthrown and the society breaks up into a number of smaller harems.

# CHAPTER 16

# The Unhappy Life of a Pasha

*The Harem*

Human males have always been fascinated by the idea of a harem. This fact alone proves that men are not by nature monogamous. However, the only men actually able to maintain a harem are very wealthy men living in poor countries where most of the people cannot get enough to eat.

People have many misconceptions about the life an animal pasha leads. They assume that all rulers of harems collect females, treat them like slaves, and from time to time select one of them to mate with. In reality, the life of a pasha varies greatly from species to species.

For example, let us examine life in a fur seal colony along the coast of the Pribilof Islands, in the Bering Sea between Alaska and Siberia. The beach, which is a little more than half a mile in length, resounds with the bellowing of more than five hundred bull fur seals. Their war cries even drown out the noise of the breakers. For two or three months the bulls have vainly awaited the arrival of the eight thousand females. While they wait, they fight, staking out territories between twenty-five and thirty-five square yards in area. The five-hundred-pound bulls conquer and defend their "beach castles," biting each other until they bleed. They hardly dare to leave their little territories even briefly to catch fish.

Finally, on the thirteenth or fourteenth of June, the first females ride the surf toward shore. At once the war cries of the males turn into tender calls. The bulls whose territories lie high up on the shore bellow enviously at those whose strongholds line the beach. Meanwhile, those along the shore try to lure the females to them with siren songs.

In Grzimek's *Animal Life Encyclopedia,* Alwin Pedersen and Herbert Wendt have written an incomparable description of what happens next: "Once a female has been captured, the bulls treat her very differently. A bull calls in flattering and cajoling tones until he has cut off the female so that she cannot return to the water. Then he grabs her by the neck and drags her to his territory on the beach. Next he starts wooing another female. But before he succeeds in trapping her, one of his rivals lures her into his territory, only to have her stolen by another bull at his rear. As the bulls battle over the females, both bulls and cows receive severe wounds. Sometimes two bulls will both sink their teeth into a female—one into her neck, the other her hindquarters—and engage in a tug-of-war. The females behave quite passively. They appear not to care which of the males they belong to."[1]

The "rape of the Sabines" continues for several days until thousands of females have been taken prisoner on shore. Finally each of the bulls in the favored positions near shore has assembled a harem of between fourteen and twenty cows, which he must guard carefully to keep them from running away. The bulls whose territories lie farther inland must be content with only four or five cows.

Their two to three months of waiting, fasting, and fighting on shore completely exhaust the bulls. During this period, they frequently lose sixty pounds of muscle and fat. The effort of mating with the females exhausts them further.

Strong young bachelor males inhabit a nearby beach. These young males constantly patrol the colony of five hundred pashas with their eight thousand females, looking for a bull too weak and exhausted to defend his territory. Sometimes a bachelor succeeds in driving away a weak older bull and takes over his territory. Thus every harem changes masters several times each season.

A bull fur seal cannot remain a pasha for more than three years, when he is at his strongest. Even then, he can only hold his territory for brief periods. After three years as pasha, he spends his life futilely attempting to make a "comeback." Each year, forty per cent of the seven- to seventeen-year-old bulls die of the bites they receive from other bulls. Only a few bulls ever reach the age of twenty-four. The high mortality rate among male fur seals explains the high proportion of females to males, eight thousand to five hundred. Thus rulers of fur seal harems lead arduous and unhappy lives. Their lot is by no means to be envied.

Male sea lions are less aggressive than male fur seals. Nevertheless, their life is not a happy one. Unlike female fur seals, which are treated very roughly, female sea lions choose their mates as soon as they arrive on shore. As a rule they voluntarily remain with the bull they have chosen.

It would be inaccurate to describe the fur seal harem as a form of polygamy, for no personal bond exists between males and females (nor, for that matter, between females and other females). No kind of marriage, not even polygamous marriage, exists here, for to the male, one female is exactly like another.

One has the impression that female fur seals come to shore not because they are driven by sexual desire, but only because they must give birth to their young on land. Nature has arranged for the mating season to coincide with the time the females give birth. Female fur seals carry their young for an entire year less two or three days. As soon as the females arrive on shore and are forced into a harem, they give birth. Immediately afterwards, the males mate with the females.

Female sea lions also bear their young as soon as they come on land and then mate with their pashas. However, the relation-

ship between male and female sea lions is very different from
that between male and female fur seals. In return for selecting
a certain bull as her pasha, a female sea lion expects to be
treated with kindness and given special attention.

No bull sea lion can mate with any of his cows until he has
courted her. Even though she may be only one of twenty cows
in the harem, she expects to be pursued and caressed. The
cows are very friendly to each other. However, to all the cows
the pasha is a somewhat awesome figure twice their size. Thus
a bull must woo a cow for at least twenty-four hours before she
begins to feel on intimate terms with him and is willing to go
swimming with him so that the two can mate.

Unlike the male fur seal, a male sea lion can sometimes
afford the luxury of going for a swim. To be sure, he must be
on guard against bachelors out to steal his harem; but the cows
themselves will not stray while he is away.

I have already stated that aggression plays a major role in
determining the form of a marriage. If the males of a species
behave very aggressively toward other males and the females
are docile, then the males will probably keep harems as male
sea lions do. The females in a sea lion harem rarely quarrel
among themselves.

Human women are more aggressive than female sea lions.
Thus women tend to quarrel among themselves and experience
feelings of rivalry more frequently than female sea lions do.
Human males are more aggressive than females. However, their
drive to band together in groups is stronger than that of women.
If men share the same basic attitudes, the bonding drive can
compensate for their higher level of aggression and prevent
them from quarrelling.

Human beings struggle to achieve a balance between aggression and the social bonding drive. However, in the area of
sexual relations, it is more difficult for women to maintain this
balance than for men. Thus women rarely consent to live together in harems unless, as in the Orient, they live in poor
countries and are seduced by the promise of wealth. They may
also consent to this arrangement if, as in ancient China, they are
raised to be the slaves of men.

Although they may sometimes be compelled to live in harems, women are by nature monogamous. Men, on the other hand, are polygamous. Nevertheless, when women acquire political power over men, they may practice polyandry like the Ngyars in Malabar. On some Polynesian islands, women who hold important positions in the tribe are permitted to have several husbands.

Sea lions do not suffer from such complex sexual problems as human beings. The females get along well together and remain faithful to their pasha as long as he does not disappoint them sexually. Fur seals must mate on the uncomfortable, rocky beach, where the bull can prevent the cow from escaping. Male and female sea lions, on the other hand, have a companionable relationship and thus are free to enjoy themselves and mate in the water.

The male and female sea lion nestle close to each other for around an hour and a half, floating almost motionless at the surface of the water. When they separate, their feeling for each other is extinguished. The bull promptly falls asleep in the water while the female hurries to land to tend her calf, which is only a few days old.

Fifteen minutes later, the pasha will probably be rudely awakened by another member of his harem demanding her rights. As the weeks pass, the worn-out bull continues to lose weight. Finally he becomes so exhausted that he falls asleep right in the middle of mating, and the disappointed female grunts with rage. Soon the other members of the harem join in her song of protest. As a rule, this mass protest signals one of the males from the bachelors' section of the beach to hurry over and fight the exhausted bull for the possession of his harem. Usually the aggressor wins.

Thus each mating season, a sea lion harem changes owners several times. As long as he is still strong and potent, the females will remain faithful to a bull; but when he weakens, they call for a replacement. Thus a bull sea lion is not really the master of his harem. He can easily be driven away if he fails to live up to his responsibilities.

Sea lions are incapable of forming permanent marital ties.

Anubis baboons

For this reason, a sea lion society never develops beyond an elementary level.

In my book *The Mysterious Senses of Animals*,[2] I discuss various animal societies in which the males keep harems. These societies include chickens in a barnyard, dwarf cichlid fish, crocodiles, zebras, and North American prairie dogs. No society based on the harem system ever develops beyond a primitive level. The germ of all more complex social orders is a family unit that remains together even after the young are grown.

In contemporary human society, the value of the traditional family unit has been challenged. People are experimenting with "open marriage," the frequent exchange of partners, group sex, and free love within a commune. As social units, all such "open" relationships are of relatively short duration. For example, as a rule young people spend more time living in communes than they would normally have spent in bachelor quarters.

Sociologists are sharply divided as to what forces play the major role in shaping a society. Some believe that we are dominated by our instincts or shaped by our environment; others

Geladas

Hamadryas baboons

believe that human reason and education are the primary architects of the social order. Which group is right?

Scientists now know what forces determine the form of a baboon society. In 1968, a detailed investigation was conducted to determine the influence of heredity, environment, and education on the social and sexual behavior of baboons.

In the wild, some male baboons keep harems. However, males and females of a closely related species practice free love. Why does a given species of baboon choose one or the other of these two types of society? Is heredity or environment responsible, or is it rather what some baboons teach others in their group? Hans Kummer set out to investigate this fascinating problem.[3]

Anubis baboons, which are native to the East African nation of Kenya, live in troops containing between thirty and eighty members. During mating season, a female baboon changes mates almost every day. She follows a certain pattern in selecting her mates. For the first few days of mating season, she offers herself to males that are low in status. Gradually she selects males that are higher in status until, at the climax of the season, she courts the leader of the troop. Probably she actually conceives her young by the leader. The lowlier members of the troop may simply serve to "train" her and help to synchronize her physically and emotionally so that she is prepared for conception.

Members of anubis baboon troops tend to form cliques or clubs—groups of bachelor males, groups of females, or groups composed of both males and females. These groups wield varying degrees of power within the troop. The baboons freely choose to join a certain "club," and are on friendly terms with other members. When the troop must travel through a danger zone, they range themselves in military formation, with guards posted at the front and rear, guards to protect mothers and children, and a main fighting contingent. As a rule, the troop is led by a "council of wise men," a triumvirate composed of older, experienced males.

Geladas and hamadryas baboons, both close relatives of anubis baboons, live in Ethiopia and Somalia. Their social order differs markedly from that of anubis baboons. The males keep small harems of two or three females that are often accompanied by a "deputy pasha." The females are not allowed to practice free love. Any female that strays too far from the harem is soundly thrashed and brought back by her lord and master.

Each harem of females lives alone with the pasha, moving around in obedience to his will. However, at night several hundred baboons gather together to sleep on the rocks at their traditional sleeping grounds. The custom of sleeping together at night does not imply that the baboons belong to a complex social unit, for when the night is over, all the pashas go their separate ways, accompanied by the females.

It is advantageous for geladas and hamadryas baboons to live in harems. Geladas inhabit the cold, barren mountains of Ethiopia, hamadryas baboons the dry desert of Somalia. Harems develop in regions where there is a scanty food supply. In such regions, small groups under the protection of a powerful male must forage over a large territory in order to find enough to eat.

A single large troop of animals would die under these conditions. Large troops can subsist only on relatively fertile plains and savannas. On the plains, anubis baboons are compelled to band together for protection against prides of lions, packs of hyenas and wild dogs, and leopards and cheetahs. A small harem could not survive here.

Clearly it is advantageous for some baboon species to live in harems and for others to live in large troops and practice free love. However, baboons have no insight into the socioeconomic advantages of their various life styles. Therefore, what motivates each species to practice its own particular life style?

Male geladas and hamadryas baboons keep their wives faithful by beating and biting them, whereas male anubis baboons are quite content to allow all the females to mate with whatever males they choose. Does this mean that the males of each group teach the females how to function in their respective societies? If so, a male hamadryas baboon should be able to teach a female anubis baboon what constitutes proper behavior for a member of his harem.

Hans Kummer decided to conduct a test to determine whether or not the behavior of male baboons in fact controls the behavior of the females. Kummer captured several female anubis baboons that had grown up in sexual freedom and released them in the territory of the hamadryas baboons. The male hamadryas baboons took the female anubis baboons into their harems and bullied them until they began to behave "correctly." By the same token, when Kummer introduced female hamadryas baboons into a troop of anubis baboons, the females quickly adjusted to the mores of their new home. Thus the female baboons are capable of living in both types of baboon society, and their behavior is controlled by the males.

Male hamadryas baboons regard the females as property. Unlike male fur seals, a male hamadryas baboon will never attempt to steal females from another male's harem. That is, he will not steal a female if he recognizes the male as one which sleeps at the same traditional sleeping grounds as himself.

To investigate the "sexual morality" of male baboons, Hans Kummer conducted a series of experiments. In each experiment, he placed a male hamadryas baboon in a cage with a strange female and allowed a second male to observe their behavior through the bars of his cage. The two males had always shared the same sleeping quarters. After several days, the second male was allowed to join the couple in the neighboring cage. Even if he was the stronger of the two males, the newcomer did not attempt to fight the first male for possession of the female. Instead he appeared very inhibited, turning his back to the couple as if he were undergoing an emotional conflict and had decided that he preferred to ignore the situation.

However, if the two males had not previously shared the same sleeping quarters, the results of the experiment were quite different. In this case, the behavior of the second male was far less honorable. When he entered the cage, he would attack the other male and steal the female from him. A male hamadryas baboon refrains from attacking another male only if he recognizes him as a trusted member of his own group. Thus in a hamadryas baboon society, it is group solidarity that guarantees the stability of the harem.

How do young male baboons go about establishing their own harems? Two courses are open to them. First, a young male may attempt to steal young females, the daughters of females in other harems. When their daughters begin to grow up, mother baboons treat them harshly, reject them, and punish them for trivial offenses. At this time, a young male may be able to remove the unhappy young female from the harem. Then, although he is only slightly older than she is, the male will "mother" his new mate, refraining from making any sexual advances. By this generous behavior, he satisfies the young female's frustrated desire for maternal love, gives her a feeling of security, and establishes with her a bond of mutual sympathy. However, the male's benevolent mood does not last. Soon he establishes a sexual relationship with his mate and begins to add more females to his harem. Then instead of maternal caresses, he administers blows whenever the females try to stray.

Human families resemble baboon families in that parents become stricter as their children grow older. They do this instinctively, as a way of helping the children to grow up and become independent. To varying degrees, all animal mothers help to force their young out into the world. For example, mother crocodiles are kind to their young on the day they hatch, but on the second day the mothers will eat the young crocodiles if they are still anywhere in sight. Mother baboons are much more patient. It takes years for young anubis baboons to become independent, and even then, young adults are not totally rejected but are allowed to remain in the troop.

It is natural for human beings to mature slowly and to remain in touch with their parents. However, when there is great conflict between parents and children, the children may band together with their peers like young animals that have been rejected by their families.. Like zebras, they may join groups composed of young people; or a young man and woman may live together like two hamadryas baboons. As a rule, young men and women who live together are seeking not sexual satisfaction but the emotional security they did not find in their own homes. Unfortunately, if one or both of the couple are not interested in making a deep personal commitment, the relationship may fail to provide the security they desire.

If human beings failed to form lasting emotional ties, our society would become like that of frogs courting in a pond. Neither we nor, for that matter, hamadryas baboons are emotionally equipped to live without the security that comes from the formation of durable ties.

There are two methods by which a young male hamadryas baboon can acquire a harem. First, he can steal young females from other harems. Second, he can serve as a "deputy pasha" in the harem of an older male. Eventually he will inherit the harem.

As harem rulers grow older, they become less jealous and possessive. If a young male approaches him behaving like a eunich, demonstrates his subservience by turning his buttocks to the pasha, performs other submissive gestures, and "selflessly" helps him to guard the females, the older male will tolerate his presence.

Gradually the "assistant harem-keeper" takes on more and more responsibility. He explores new feeding grounds and guides the troop in searching for food. He decides when the troop will move on, where they will go, and when they will return to their sleeping rock for the night. However, he obtains no sexual privileges until the pasha dies, when he automatically inherits the care of the females and the young.

Both methods of acquiring a harem presuppose considerable foresight on the part of a young male baboon. Baboons do in fact possess a high degree of foresight. In the evening, around five hundred hamadryas baboons gather at their sleeping rock for the night. At this time, some of the younger baboons play a game, throwing stones down the steep cliffs trying to make as loud a noise as possible. Once a scientist observed a young baboon on a cliff more than thirty feet above the spot where five baby baboons were playing, unaware that they were in danger.[4] The young baboon was toying with a heavy stone and obviously was longing to throw it down the cliff; but he held back until the five youngsters had moved some distance away. Incidents like this demonstrate that baboons are able to foresee the results of their actions.

It is often claimed that the invention of weapons like the sling and the spear made human beings more aggressive. In theory, such weapons enabled a man to kill other men without coming into close personal contact with his victims, and thus reduced his inhibition against destroying his own kind. That is, it counteracted an innate inhibitory mechanism (IIM). Yet a young baboon is capable of realizing the destructive power wielded by a long-distance weapon like a stone. Moreover, he is capable of choosing not to exercise that power.

Assuming that baboons possess a high degree of insight, do they possess sufficient insight to deliberately adapt their social structure to environmental conditions? Investigations of baboons living in the border region between the territories of the hamadryas baboons and the anubis baboons supplied the answer.

Food is scarce in the twelve-mile region. If the anubis baboons in this region could consciously choose the most advantageous social order, they would cease living in a large troop and the males would keep small harems like the hamadryas baboons. However, this is not what they have done.

To be sure, during the day the anubis baboons break into small groups that forage for food. However, each day the groups are composed of different individuals. One day three anubis baboons known as Ernie, Bert, and Crumb went out foraging together. Ernie and Crumb quarrelled, so the next day Crumb travelled with another group, and Bibo took his place with Ernie and Bert. Sometimes groups of females travel alone.

Despite the scanty food supply, anubis baboons in the border region continue to practice free love. In other words, despite the fact that the environment does not favor their social order, the male anubis baboons are unable to alter their behavior to suit circumstances. Thus none of the males attempt to form a harem.

In 1970, three baboon sleeping communities totalling one hundred and eighty baboons occupied the border region. These baboons were hybrids, a cross between anubis and hamadryas baboons. Young male hamadryas baboons had stolen young female anubis baboons for their harems, and these unions produced hybrid offspring. Naturally, male anubis baboons made no attempt to seize female hamadryas baboons.

The hybrid offspring of the mixed harems sometimes mated with anubis baboons, sometimes with hamadryas baboons, and sometimes with other hybrids. In all three cases, they produced hybrid young. On the surface it appeared that the hybrid baboons practiced the harem system. However, observers discovered that every few days or weeks, the females changed harems. To be sure, some pashas attempted to hold onto the females after mating season, but they were no longer as aggressive as the pure male hamadryas baboons, and the females invariably escaped them sooner or later.

The hybrid baboons exhibited a wide spectrum of sexual behavior, from the maintenance of temporarily stable harems to extremely brief pair formations between a single male and female baboon. The hybrid males, which had little skill in guarding females, never succeeded in keeping more than one female at a time in their harems. However, the fact that a pasha might possess only one wife did not mean that their relationship was monogamous. The small size of his harem resulted solely from the male's inability to control more than one female at a time.

The sexual behavior of hybrid male baboons varied with

their heredity. That is, their behavior depended on whether hamadryas or anubis baboon traits predominated in their genetic makeup. Since the females were capable of adapting to either life style, the genetic material inherited from the mother did not predispose the hybrid males to keep harems.

Baboon behavior shows that a baboon society is shaped by three factors—heredity, environment, and education. The males inherit the technique by which they deal with females. Then they teach the females whether or not they must live in harems. Finally, the environment favors the development of a certain type of social organization.

The forces that shape human societies are more complex and variable than those that shape the social order of baboons. However, we too are influenced by heredity, environment, and education. When these forces harmonize—e.g., if our education does not conflict with inherited tendencies—our societies are peaceful. When there is conflict between heredity, environment, and education, the result is oppression and revolution.

Our social theoreticians, both conservatives and revolutionaries, are too one-sided in their views of our social structures. If fanatics attempt to educate people to live in a way contrary to the natural order (i.e., if education conflicts with heredity), or if they ignore the economic and technological demands of a changing environment, the result will be the destruction of mankind.

# PART SEVEN

## Forms of Marriage

# CHAPTER 17

# Marriage Partners Who Do Not Know Each Other

*"Local Marriage"*

If we define marriage as a relationship between male and female based on the personal attachment of the partners, which transcends sexual needs and persists for some time after sexual intercourse occurs, then animals enter into various sexual relationships that cannot be classified as marriage. One such relationship is that between a pasha and the females in his harem.

Besides the harem, a second form of non-marriage is prevalent in the animal kingdom. Sometimes male and female are married not to each other, but to a place—the nest or the hunting territory. The partners are personally indifferent to each other, and neither cares if its mate dies or disappears and is replaced by another. In some cases the two mates may not even be able to distinguish each other from other members of their species. Thus when they are away from the nest, they behave like total strangers. Various species of insects, fish, reptiles, birds, and mammals engage in this kind of sexual relationship.

A species of dragonfly known as *Calopteryx haemorrhoi-dalis* spends part of its adult life sitting in large groups with other dragonflies. The rest of the time it lives and hunts alone. The insect spends its "social hours" in a bush or tree on the bank of a river or a stream. On almost every leaf a dragonfly sits sunning itself, taking shelter from the wind. Just as all Moslems face toward Mecca when they pray, all dragonflies face toward the sun. Although males and females are in close proximity, they do not mate. Both sex and fighting are taboo at these meeting-places.

Provided the weather is fine, the sexually mature dragonflies leave the meeting-place at daybreak. Then the males fly up- or downstream looking for a private territory along the bank. Each male stakes out a territory several yards in length and extending about a yard inland from the bank of the stream. The land farther inland constitutes a neutral zone.

The male dragonfly marks his territory by flying around its borders so that other males will not intrude. Now he behaves very differently from the way he behaved when all the dragonflies were sunning themselves together. When he was socializing, for hours he sat beside other males and ignored all the females. Now he attacks any male that dares to enter his territory and passionately courts all the females that approach him. Thus these male dragonflies behave much like male wildebeests and gazelles. When they live in a group, they are peaceable and sexually impotent. However, the possession of a territory turns them into fierce fighters and passionate lovers.

The male and female dragonflies recognize each other by the flight patterns. Males fly in a wavy pattern and move faster than the females, which always fly in a straight line.

In the evening or when the weather turns bad, both males and females leave the males' territories and fly back to the meeting-places, where the not-yet-fully-matured dragonflies have been waiting all day. When travelling to and from their territories, the insects fly through the neutral zone. During the flight, both males and females revert to being emotionally sexless beings. The males lose interest in the females and abruptly cease to regard the other males as rivals.

This behavior is an example of intermittent social bonding. The periodic struggle for territory and the resultant emergence of the aggressive and sexual drives constitute asocial behavior that destroys the social bond among the dragonflies. Aggression, sexuality, and the drive to possess territory are reproductive drives inimical to the social bonding instinct that causes the insects to assemble at common meeting-places.

Let me repeat: The male and female dragonflies can mate only in the male's territory, not at meeting-places or in neutral zones. Within the borders of the male's territory, both sexual partners are indifferent as to which insect they mate with. The male courts every female that comes along. The female is primarily interested in finding a suitable place to lay her eggs. The males are careful to select territories where the depth of the water and the speed of the current are such that they will not damage the eggs, and where there are abundant places for the female to hide while she lays them. In other words, it is a *place*

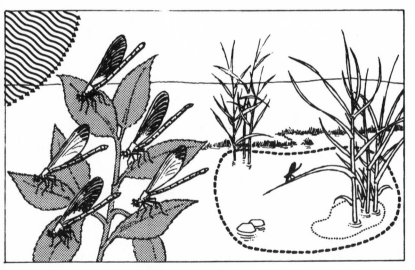

On the left, male dragonflies are sitting close together. As long as they are in a group, their behavior is peaceful and asexual. A few yards away, a solitary male dragonfly aggressively defends his territory against other males and mates with all the females that come near.

that interests the two sexual partners and that brings them together for their brief encounter.

Blind goby fish also marry a place rather than another fish.
These tiny creatures, which are little more than one inch long,
live in small caves along the Pacific coast of southern California.
The caves in the stone that gobies favor as homes are usually
plugged up with sand or silt, and the gobies cannot dig. Thus
to find shelter, the fish must rely on the aid of a creature that
can dig and that will tolerate the presence of a "tenant" in its
house.

The ideal housemate of the blind goby is the little tropical
mole crab. Sensory cells along the fish's sides enable it to measure the vibrations of objects in its environment. These cells and
its sense of smell enable the blind fish to locate a tropical mole
crab. When it finds a crab with a "room to let," the goby's
troubles are over. The fish lives on microscopic animals that
stream into the cave while the crab is engaged in its never-
ending excavations to keep the cave free of sand. However,
this symbiotic relationship has one disadvantage. If the crab
dies, the fish will die too, unless it can find another crab very
quickly.

Despite their size, blind gobies are very aggressive. The
males will not share their homes with other male gobies. However, the males do not attack females, nor do the equally aggressive females attack males. Thus sometimes a male and female
blind goby will end up living in the same cave. Both fish come
to the cave seeking food and shelter, not love. Nevertheless,
eventually the two of them mate. They have no choice, for they
are the only two fish of their species in the cave. The first
"tenant" in the cave must accept as a mate the first goby of the
opposite sex that enters.

Scientists have conducted experiments on blind gobies,
removing the female or the male and replacing it with another
fish of the same sex. The other fish does not care if its mate is
replaced by a stranger. The fish are married to the cave, not to
each other.

Green lizards also marry their homes rather than other
lizards. Each partner drives away all members of the same sex

which try to enter its territory. However, neither sex attacks members of the opposite sex. No personal bond exists between sexual partners. However, the inhibition against attacking members of the opposite sex, in conjunction with the dependence of both partners on a common shelter, replaces the bonding instinct and enables the mates to establish a relationship.

At first glance it may appear that blind gobies and green lizards both practice monogamy. Both species are highly aggressive. In each case, the male drives away all other males that enter his territory, and the female drives away all other females. Thus neither species can practice polygamy or polyandry. Nevertheless, blind gobies and green lizards are not genuinely monogamous, for in both cases the mates are quite indifferent to each other, and neither mate cares if the other is replaced by another animal of the same sex.

Because this type of sexual relationship closely resembles monogamy, people have mistakenly assumed that storks are monogamous birds. In the popular imagination, storks symbolize marital fidelity and familial happiness. In reality, storks are married to the nest and not to each other.

Every spring, many storks arrive in the town of Bergenhusen in Schleswig-Holstein, Germany. The birds have travelled all the way from Africa. The males arrive before the females, and at once each male returns to his nest of the previous year. A few days later the females fly into town. They too look for their old nests. Thus it is their fidelity to the nest that brings the two partners together.

One day Peter, a male stork, returned to his nest near the church.[1] Later a female arrived at the nest, and the two of them greeted each other by clattering their beaks and walking around ceremoniously in a circle. The birds behaved like old friends meeting again after a long separation. Observers assumed that the female must be Luise, Peter's mate of the previous year. However, the female's identification band revealed that she was really Renate, a four-year-old female that had never had a mate and that was native to the neighboring town of Börm.

Peter did not seem in the least disturbed by the fact that the female in his nest was not his mate Luise. Two days later,

Peter and Renate were in the middle of a courtship duet, clattering their beaks and bending backward until the crowns of their heads touched their backs. Suddenly Luise landed at the edge of the nest and began to stab at Renate with her beak. Then the females fought a ferocious battle that might easily have proved fatal to one of them.

Peter seemed quite indifferent to the fray. Making sure that he did not get caught in the middle, he walked away along the coping of the roof, turned his back on the combatants, and gazed intently into the foliage of a nearby tree.

Twelve minutes later the battle was over. Luise had driven away Renate and established her right to occupy her old nest. At once Peter came over and began to court Luise as he had been courting Renate. He would have accepted as his mate any female that happened to be in the nest. By the same token, Luise would not have bothered to defend Peter if his right to the nest had been challenged by another male. Moreover, if Peter had been driven away, she would not have left the nest with him to look for another home, as a greylag goose, jackdaw, raven, or human mate would have done.

The behavior of storks when they meet and greet each other reveals that they are in fact able to recognize each other. In other words, Peter knew perfectly well that the female that first landed in his nest was not his former mate Luise. Nevertheless, no personal bond exists between stork mates. If storks are faithful, they are faithful only to the nest. When a purely sexual tie exists between mates, neither mate cares if the other is replaced by another animal of the same sex. Sexuality alone cannot create a lasting tie.

The ability to distinguish between individuals of one's own species is a prerequisite to the formation of personal ties. Nevertheless, the behavior of storks reveals that the mere recognition of another individual is not sufficient to create a relationship.

The sexual relations of skylarks represent a transitional phase between "local marriage" and genuine monogamy.

The migratory skylark leaves its winter quarters and returns to Central Europe in the middle of February, when it is still winter. The males occupy territories ranging in area from fifty-five hundred to sixteen thousand square yards. The females arrive at the breeding grounds around ten days after the males and occupy territories approximately equal in size to those of the males. As a rule the territories of the males coincide with those of the females. The occupation of the same territory by a male and a female is the prerequisite for the formation of "local marriage."

The female skylark makes the first advances by fluttering over to the male. However, frightened by her own daring, she promptly retreats.

In any type of sexual relationship, both partners must overcome feelings of fear and aggression. The blind goby and the green lizard instinctively refrain from attacking members of the opposite sex. Male and female storks clatter their beaks and make other friendly gestures to assure each other of their good intentions. However, skylarks have a more difficult time in establishing a relationship. When the male observes the shy female, he begins to court her. Initially, his courtship merely serves to frighten her away. When she flees, the male pursues her, trying to drive her back to the center of his territory.

Up to this point, there is nothing unusual in the behavior of skylarks. But when the female, fleeing the male, enters the territory of two other larks that have already mated, something quite remarkable occurs. The male lark, afraid of intruding into the other male's territory, gives up the chase and begins to call the female in seductive tones. Meanwhile, the female neighbor, angered by the intrusion of another female into her territory, promptly attacks her. Then her mate joins in the attack, trying to protect her from the "intruder." The two mates drive the female lark back across the border and into the waiting wings of the male that is courting her. That is, they make sure that she is faithful to her "marriage."

Although male and female skylarks are faithful to a place, they form a personal tie transcending the tie to their territory. When the two neighboring skylark mates drive the female back into her own domain, the male bird takes her side in the fight. Thus his behavior differs from that of the male stork, which stands by passively while two females fight, indifferent to the outcome of the battle.

However, skylarks do not make devoted mates. One day a male lark named Till was attacked by another male that was trying to take over Till's territory.[2] The two birds fought breast to breast several yards above the ground, pecking, digging their claws into each other, and beating each other with their wings. Then the two of them plummeted downward. Catching themselves just before they hit the ground, they flew into the air for another round. They continued to fight for ten minutes until Till finally drove away his rival. Throughout this time, Till's mate Ariane made no effort to help him. Thus the male seems more concerned about the welfare of the female than the female does about that of the male.

After Till and Ariane had reared their first brood, the male owner of a neighboring territory was killed by a condor. Till, noticing that the territory was now unoccupied, promptly claimed it for himself. Every fifteen minutes or so he flew one hundred and fifty to two hundred and fifty feet into the sky, singing to announce his ownership. Naturally, he also consoled the bereaved widow. This is an example of territorial bigamy. Clearly, to male animals the grass always looks greener on the other side of the fence.

Till spent so much time with his second mate that Ariane did not lay any more eggs and thus failed to raise her second annual brood. However, the following year, Till became "monogamous" again and lived on his old territory with the neglected Ariane.

The personal bond between a male and female skylark is comparatively weak. It is never stronger than the tie to a particular territory. Juan D. Delius observed thirty pairs of skylark mates along the coast of northwestern England on the Irish Sea.[3] The birds lived in the national park of Ravenglass. Sixteen of the thirty skylark couples stayed together for two mating seasons in succession, and in all these cases the birds returned to their own territory. In five cases the females changed territories. In three cases the males changed territories because they were driven away by other males. The new male occupants of these three territories mated with the females that had occupied the territories the previous year. In six cases, both the males and the females moved to other territories. In each of these cases, the males and females moved to separate territories and mated with new mates. Thus territory is always the basis of a skylark marriage. Skylarks never mate because a personal tie exists between the partners.

In various types of marriage, the tie to a particular place may play a role in establishing a bond between two sexual partners. Human beings have a feeling of "being at home" when they are in a familiar place. The attachment to home profoundly affects the course of many marriages.

# CHAPTER 18

# Only Aggressive Mates Stay Together

*Seasonal Mating*

Seasonal mating is a form of monogamy that cannot for very long withstand the pressures of the environment. One animal that engages in seasonal mating is the marsupial known as the Tasmanian devil. The devil is a predator about thirty inches long that devours anything, living or dead, that is not strong enough to resist it. Tasmanian devils are solitary animals that live and hunt alone in the bush country on the island of Tasmania south of Australia. The devil will not permit competitors to enter its large territory, and the male drives away females of his species as well as other males.

Shortly before mating season in April, when the Tasmanian summer is almost over, the male's behavior alters. On bright moonlit nights he dashes across the hilly plain looking for a female. When he finds one, he does not court her but behaves like a sheepdog that has cornered a stray sheep. Snarling, gnashing his teeth, and biting the female, the devil drives her into his territory and then into the cave in the rocks where he lives.

Male and female do not mate until two weeks after the female is captured. Not once during the two weeks does the male permit the female to poke her head out of the cave. He guards her as if she were a dangerous criminal. Whenever the female tries to escape, he arches his back, spits, snarls, bites, and foams with rage, trying hard to live up to his name.

When she is first captured, the female devil is not yet sexually mature and ready for conception. This is why the male and female observe two weeks of continence before mating. Fear and enforced captivity cause the female's sexual organs to mature rapidly. Thus the male's "diabolical" behavior helps to synchronize her sexual drive with his own.

Shortly after they mate, the physically smaller and weaker female turns the tables on the male and begins to bully him unmercifully. Once she is pregnant, she becomes even more aggressive than he is. But despite the aggressiveness of both partners, they remain together for some time.

The young, four at most, are born at the end of May or the beginning of June. As is the case with all marsupials, the young are incredibly tiny, no more than half an inch in length. Each of the young devils holds fast to the mother by sucking one of the nipples inside her pouch. The pouch of a mother kangaroo opens at the front of her body, but the pouch of a Tasmanian devil opens at the rear. If it opened in front, it would drag against the ground and scoop up earth when the mother was creeping through the reeds, and this might damage the young. Moreover, the opening is so tightly sealed that as a rule a human being can see neither the pouch nor its contents. Only when the young get to be fifteen weeks old, a tail, a leg, or a head sometimes pokes out through the slit.

When the young reach this age, it is time for the parents to dig a hole inside the cave to hold their offspring. Both parents dig the hole and line it with hay. The mother continues to nurse her young for another five months. While she goes hunting, the father guards their family. If the father did not assume this duty, enemies would enter the cave and eat the young. Guarding his family is the father's principal task until the young are grown. Thus for nine and a half months, he is forced to stay married.

At the beginning of February, the devils are old enough to become independent. They do not become sexually mature until they are two years old, but at the age of nine and a half months, their instinctual bond with their parents and their brothers and sisters disappears. Each of the young devils enters the big dangerous world all alone.

When the young set out on their own, the adult male and female devils quarrel and separate. Each lives alone for two and a half months, until the next mating season, when the male goes hunting for another female. Chances are that he will not find his former mate, but a completely strange female that he must imprison for two weeks. Then the whole drama begins again.

The marriage of a Tasmanian devil lasts only as long as the male is needed to help care for the young. The longer it takes for the young of a species to reach adulthood, and the longer the father's help is needed in raising the young, the longer an

When the female Tasmanian devil tries to escape her mate, he behaves as if he were trying to live up to his name.

animal marriage will last. However, the reverse is not always true. Some animal mates that drive away their young soon after birth nevertheless remain faithful to each other all their lives. In fact, animals of some monogamous species that mate for life pay no attention whatever to their young. The emerald cuckoo, the parasitic weaver bird, and the black-headed duck all lay their eggs in the nests of birds of other species and allow these foster parents to raise their young.

After laying and fertilizing their eggs, butterfly fish do nothing further for their young; yet these fish mate for life. Probably they are monogamous because they are a very rare species. By remaining together, the male and female are spared the time and trouble involved in seeking a new sexual partner each mating season, and thus have a better chance of reproducing.

Clearly, animal parents do not remain together solely for the sake of their offspring. However, it is true that when the mother cares for the young alone, the marital tie is severed almost immediately after mating takes place. In these cases, a "seasonal marriage" consists of courtship, mating, a honeymoon lasting only a few hours or days, and a divorce. Such marriages are so brief that they do not really constitute marriages at all. Male and female porcupines, golden hamsters, tigers, and polar bears all conduct this kind of brief liaison.

Polar bears are among the largest bears in the world. When standing on their hind legs they are more than eight feet tall, and they weigh almost fourteen hundred pounds. These tall, muscular creatures are by nature solitary and unsociable. They wander alone through a region the size of Europe, driving away all other polar bears, both male and female.

Bears of all species are solitary and unsociable. Their unsociable character is revealed in their faces, which are quite expressionless. A bear's cheeks and forehead contain no muscles to alter the animal's expression and reflect its emotions. We cannot tell by looking at a bear whether it is feeling angry, afraid, or friendly. Even after years of working with a polar bear, an animal trainer cannot tell what his bear is going to do next.

Animals change their facial expression in order to communicate their feelings to members of their own species. Solitary animals like the polar bear do not need to have expressive faces. However, they are caught in a kind of vicious circle: Solitary animals do not need expressive faces, but unless they develop expressive faces, they must remain solitary. Animals like the polar bear, which cannot communicate their feelings to their own kind, have no way of establishing a close bond with a sexual partner, and thus cannot enter into a durable marriage.

When human beings look expressionless, it is not because they lack the muscles necessary to communicate their emotions. They may be emotionally cold, egotistical, insecure, or have bland personalities; or they may simply be very good at controlling their feelings.

Obviously, a polar bear's inability to communicate its feelings to other polar bears proves a great handicap when the bear is looking for a mate. When the male finds a female, both of them seem uncertain whether or not they should attack each other.

Then the two polar bears begin to chase each other around. As a rule they end up fighting. Standing on their hind legs, they grapple with each other like two wrestlers, holding each other with their forepaws and pushing and shoving with all their might. However, they do not bite each other, for each instinctively refrains from inflicting serious injury on a sexual partner. Gradually their pawing and mauling turns into caresses. Through physical contact, the two bears manage to communicate the friendly intentions they cannot express with their faces.

A polar bear romance lasts for two to three weeks at most.

Giraffes are the very soul of friendliness and tranquillity. However, these "walking lookout-towers," which may reach a height of almost twenty feet, never marry. The males and females separate immediately after mating.

Before they mate, brown bears growl at each other and fight. Gradually their blows turn into caresses. However, a brown bear marriage lasts for only two or three weeks. Afterwards both bears go back to living alone.

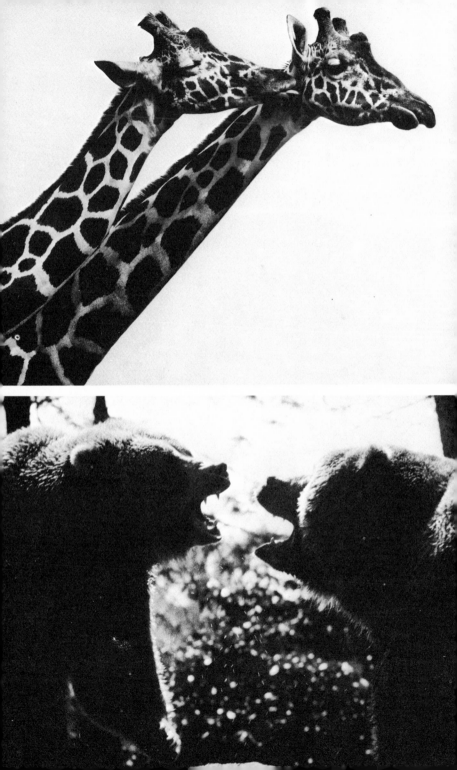

As their passion cools, the mates treat each other more roughly and finally separate. The chances are that they will never see each other again and that during the next mating season, both of them will find new mates.

Compared to the polar bear, the giraffe seems a peace-loving, amiable individual. Nevertheless, the sexual relations of these walking lookout towers, which range from twelve to twenty feet tall, are even more transient than those of the polar bear.

Giraffes live in loosely organized, rather anarchic herds. Whenever they choose, males and females change herds. The only rule is that bulls must always live in herds with other bulls and cows with other cows. The two sexes display no interest in each other, except that the bulls never let the herds of females get completely out of sight.

There is a ranking order in giraffe herds. As a rule, the tallest giraffe is the highest in rank. The highest-ranking giraffe always carries its head high. Lower-ranking giraffes must lower their heads slightly and thrust them forward when they pass a higher-ranking member of the herd.

Mothers and children are rarely seen together. Female giraffes care for their young for only one or two months and then become indifferent to their fate. Usually the young giraffes manage to trot along after the herd. If, while several herds are drinking at a water hole, a young giraffe decides to change herds, its mother does not even appear to notice.

Giraffes are peace-loving creatures that do not experience strong emotional ties. Thus they devote little time to romance, either before or after mating. Occasionally one of the bulls strolls over to a female herd to smell the females and find out whether any of them are sexually mature and ready to mate. If he finds a mature female, he does not bother to engage in a lengthy courtship, but simply mounts her and then returns to the other males.

The extreme brevity of the giraffe's marital relations is the consequence of its docile nature and of the fact that it lives in herds. A giraffe is a social animal that enjoys living in small groups with other giraffes. No giraffe cares whether another giraffe is a stranger or an old-established member of the group. That is, giraffes live in an "open society," as opposed to the closed societies of wolves, lions, and anubis baboons. Giraffes are always willing to accept new members into the herd.

Giraffes may occasionally fight among themselves. For example, if a low-ranking giraffe "forgets" to lower its head in salute when a higher-ranking giraffe passes by, the two animals will fight by swinging their long necks together as hard as they can. However, for the most part giraffes are not at all aggressive, form no hostile cliques or sub-groups, and are never afraid of each other. This means that when a male and female giraffe meet, neither animal is particularly excited by the encounter, and neither animal has to struggle to overcome its fear and aggression. When male and female are so indifferent to each other, there is no need for the male to engage in an elaborate courtship. Moreover, in a herd there is no need for animals to form close personal ties. Thus giraffes mate quickly and just as quickly forget about each other.

In the animal kingdom, peace-loving animals that live in herds and do not fear each other never establish lasting marriages. Such animals enter into a "seasonal marriage" of short duration.

Among animals, the psychological basis for a lasting marriage is the conflict between a personal bond on the one hand and fear and aggression on the other. When male and female must struggle and take risks to overcome their own and each other's resistance to a close relationship, the relationship can persist beyond the sexual act. The durability of such a relationship depends on how long the personal tie between the mates can effectively counteract their fear and aggression.

Male and female giraffes do not experience the drive to forge a personal bond, for they are so gentle that they do not need such a drive. The tigress, on the other hand, must work to establish a personal bond with her mate. However, she ceases to feel this bond the moment her sexual desire is extinguished. Thus different as giraffes and tigers are, both engage in extremely brief "seasonal marriages." Both polar bears and Tasmanian devils are highly aggressive. Their behavior demon-

A five-year-old female pied flycatcher shows her one-year-old mate the kind of insects he should bring their offspring. Being so young, the male has had no experience in caring for baby birds.

strates that once the bonding instinct has established peace between two mates, it can hold them together for an extended period.

If it is biologically advantageous to a species for the male and female to stay together until the next mating season, they may engage in a permanent monogamous marriage. This form of marriage spares them the trouble and risk involved in searching for a new partner every year.

Later I will discuss the permanent monogamous marriage in detail. For now, let us return to the metamorphoses of the "seasonal marriage."

The pied flycatcher, a member of the subfamily of flycatchers, finds a new mate every spring. Pied flycatchers prefer to mate with birds their own age. One-year-olds almost always mate with one-year-olds, and so forth.

The behavior and the appearance of these birds reveals their age. The older a male is, the darker the feathers on his back are during mating season. The older a female is, the earlier she leaves the Sudan to fly home to central Europe in the spring. The older females lay their eggs earlier than the younger females and hatch a larger brood of young.

Clearly it is advantageous for flycatchers to mate with birds of the same age. The sexual development of the two birds will be approximately the same. Thus it will be relatively easy for them to synchronize their sexual maturation so that they are ready to mate at the same time. Moreover, both birds will have had equal experience in raising the young. It is extremely important that both parents know how to care for their offspring.

Once, contrary to custom, a five-year-old female flycatcher named Cheep-Cheep mated with a one-year-old male called Wuti.[1] When the young birds hatched in their nest inside a tree trunk, Wuti began bringing them big, fat caterpillars that he found in abundance in nearby bushes. The caterpillars were too large to fit into the tiny beaks. Thus the female had to try to feed the young birds alone, hunting for flies, which were much harder to capture than caterpillars. She also had to keep pushing the caterpillars out of the nest again. After two days, Wuti finally realized that he could not feed the young birds his own favorite food and began hunting for flies like his mate. However, by this time two of their seven young had already starved to death.

Naturally, the flycatchers do not deliberately choose to mate with birds of the same age; nor do they feel an aversion to birds belonging to a different age group. Instead, nature has arranged for birds of the same age to be in close proximity, so that they will inevitably mate with each other.

The younger the birds are, the greater is their thirst for adventure, the farther south they fly into the Sudan during the winter, and the later they return home in spring. First to arrive home are flycatchers that are three years old and older. Three days later the two-year-olds arrive, and three days after that the one-year-olds.

As a rule, the flycatchers quickly mate with the other birds in their vicinity. Thus each group has finished mating by the time the next group of younger birds arrives. Only older birds that have failed to find a mate will attempt to mate with birds of a younger age group.

Thus in some species, the thirst for adventure serves a useful biological function.

Like young pied flycatchers, the younger members of many animal species migrate to their breeding grounds later in the year than their older fellows. This is true of birds such as northern gannets, sea gulls, and terns, all of which raise their young in breeding colonies. Moreover, older male seals, fur seals, and walruses arrive at their colonies earlier than the younger males.

It is more difficult to establish whether older members of migratory species that do not breed in colonies also return home earlier than the younger members. To study the migratory patterns of the pied flycatcher, Rudolf Berndt[2] had to rely on the help of countless volunteers in the southeastern region of Lower Saxony. These volunteers banded one hundred and thirty pairs of pied flycatcher mates housed in special nest-cases and observed their behavior over a period of years. For the first time, a scientist succeeded in proving that older members of a species that does not breed in a colony migrate home earlier in the year than younger members.

Like young pied flycatchers, the younger members of species that breed in colonies travel farther away and return home later than the older members of the colony. They arrive at the courting or breeding grounds after the older animals have already established a harem, wooed their mates, and begotten their young. This behavior pattern is advantageous to offspring, for it prevents an inexperienced mother or father from counteracting the efforts of his or her more experienced mate.

Many animals have to learn how to care for their young. As a rule, most of a bird couple's first brood do not survive. However, the young parents soon learn not to repeat their mistakes. The eggs of one-year-old pied flycatchers hatch at a different time and in different locations from those of the older birds. Thus the novice parents cannot interfere with the efforts of the older, more experienced parents.

The love of young animals (as well as young human beings) for adventure serves to divert some of their tempestuous energy from sexual outlets into other channels. For example, for the first three years of their lives, Adelie penguins do not ever come to the breeding colonies. The older adults breed along the coast of Antarctica. After a brief honeymoon, the birds must spend six arduous and dangerous months raising their young. Meanwhile the younger penguins swim around among the ice floes or wander the southern oceans. At the age of four, they travel to the breeding colony, arriving long after their older peers. After wandering around the great marriage market where a hundred thousand penguins are on display, they realize that they have arrived too late to find mates and slip back into the ocean.

The following year, the penguins hurry to their destination. Each penguin returns to a place it has not seen for more than four years: the place where it hatched from an egg, was tended by its parents, and played with neighboring youngsters. Here it finds the marriage market already open. Older penguins are trying to assert their rights to the stony nests they occupied the year before. However, storms have devastated the shoreline, and moreover, penguins are very short-sighted. They have difficulty recognizing their own nests in an area where there are some four nests to every square yard of territory. Sometimes the owner of a nest may not arrive until two or three days after its nest has already been occupied by another penguin. Thus the penguins go on fighting, boxing each other's ears with their stumpy wings, and jabbing each other with their beaks for days on end.

To add to the turmoil, all the penguins are desperately struggling to find mates. The males and females separate when they leave the colony, and spend many months apart. However, when they return to the breeding colony, all the penguin mates that raised young together the previous year try to find each other again. Penguins are not married to a territory. Nevertheless, they tend to return to the place where they were born or to the location of their last year's nest. This fidelity to a place helps former mates to find each other again.

However, no penguin can afford to wait indefinitely for its former mate to arrive, for the mate may have been eaten by a leopard seal or a killer whale. Moreover, there are plenty of other marriage candidates to choose from. Many young four-year-old males and females are only too eager to press their suit or be courted with a stone in the manner I have previously described. (See Chapter 3.)

Among a penguin's first marital obligations is the building and furnishing of a proper nest. The female curls up on the nest while the male gathers building materials. Then the female builds a stone wall around the nest. There are countless small stones only a few hundred yards away from the mates. However, no self-respecting penguin wants to carry stones such a long distance when it is much easier just to steal a few from his neighbors.

It is amusing to watch a male penguin creep up on a female neighbor and pull out a stone from underneath her. If she turns around before he gets his booty, the thief, looking innocent, will gaze at the sky as if he were saying, "Oh, I just happen to be standing here!"

Once zoologists led a female penguin over to the nest of the penguin that had stolen many stones from her nest. She did not recognize any of the stolen property. But woe to any penguin that is caught in flagrante delicto! The owners of the rifled nest will pursue him halfway through the colony, stabbing him with their beaks and beating him with their wings.

For days the penguin metropolis is the scene of bitter quarrels and unhappy triangles, jealous tantrums and petty thievery, brawling and infidelity. But twenty-seven days after the penguins return to the colony, the females lay their eggs, and from then on, all the colonists are models of virtue and all the penguin marriages are secure. The three-year-old males and females that arrived after all the other penguins, and others that have been unable to find mates, plunge back into the water to hunt for shrimp and fish. The penguins that remain on land settle down to several months of peacefully raising their young.

Once all the furor subsides, how many Adelie penguins have managed to find their way back to their former mates?

Eighty-four per cent of the older penguins choose the mates they had the previous year. Only sixteen per cent yield to the advances of their younger peers, and they do so only if their former partner arrives many days after the younger penguins.

Fifty per cent of the younger penguins, which have raised only one brood, choose another partner, regardless of whether their old partner has arrived late. We can only assume that some of them consider themselves "unhappily married" and seize the opportunity to improve their lot the second time around.

Penguins of other species are less faithful to their mates than Adelie penguins. I have already mentioned that Adelie penguins are extremely aggressive birds. The larger king penguins have more docile dispositions. They live more quietly

On the right is the Adelie penguin. The smallest of these three penguin species, the Adelie is also the most quarrelsome; yet it makes the most faithful mate. In the center is the emperor penguin, the most peaceable but the least faithful of the three. The king penguin on the left is of medium height. It is also a bird of medium faithfulness, being more faithful than the emperor but less so than the Adelie penguin.

than Adelie penguins, quarrel less over territory, and do not bother to build stone fortresses around their nests. The male and female take turns holding their single egg on their warm feet, which replace the nest as a shelter for the egg. King penguins do not raise their young in any one place, but continually move around. Because they do not stake out territories, they do not need to be as aggressive as Adelie penguins and choose their partners with less fanfare.

When the male and female of a species are not aggressive, they do not have to struggle to overcome each other's fear and aggression so that both partners can tolerate a sexual relationship. As a result, it is easy for both male and female to form new ties each year, and it is unnecessary for them to return to their former mates. Thus king penguins do not mate with their mates of the previous year as frequently as Adelie penguins do.

Emperor penguins weigh twice as much as king penguins and are twice as clumsy. Like king penguins, they use their feet as a nest for the eggs and continually move around. Thus they do not bother to stake out territories. When Antarctic snowstorms rage across the ice for days on end, groups of from five to six hundred emperor penguins huddle together to keep each other warm. Thus emperor penguins are even more sociable than king penguins. No emperor penguin greets a member of its species with hostility and aggression.

During courtship, emperor penguins retain their tractable dispositions. Instead of squabbling and screeching, the males woo the females with gentle calls. However, emperor penguins have even less reason than king penguins to seek out their former mates in mating season. When all the males and females get along well, one partner is as agreeable as another. As a result, only fourteen and a half per cent of emperor penguins

To win a female's love, a male tern must court her with an especially beautiful fish in his beak. In bird language, holding a fish in one's beak means "I can support a family." Later when the female is busy sitting on the eggs, the male tern must feed her. The picture shows a male returning to the nest with food.

return to the mates they had the previous year, compared to eighty-four per cent of Adelie penguins.

The rule that applies to bears and giraffes also applies to penguins: Only aggressive animals are capable of marital fidelity. Thus fidelity is a feeling of closeness that springs from insecurity and fear of contact with unfamiliar, aggressive members of an animal's own species.

Emperor, king, and Adelie penguins represent three phases in the transition from seasonal to permanent marriage. The relationship between the male and female woodchat shrike (I have already discussed these birds in Chapter 11), also represents a transitional phase between temporary and more durable ties. The male and female woodchat shrike are monogamous and raise their young together. Once the young birds are fledged the adult shrikes need no longer struggle to defend their twenty-acre breeding territory against other shrikes. Now the entire family roams around hunting for food together. Soon the

These albatross mates have not seen each other for a year and a half. The male arrived at their old breeding grounds five days before the female. When they met, the two mates began a courtly ceremonial. They greeted each other by dancing around each other in circles, "fencing" with their beaks, bowing, clattering their beaks, and showing each other their nesting site. Finally they sat down quietly and caressed each other by running their beaks through each other's feathers. Albatrosses breed once every two years. For a year and a half the mates live apart, soaring over the oceans. Then they return to their breeding grounds to spend six months together.

Bull hippopotami do not fight for control of a harem; they fight for "ringside seats" around a group of from twenty to one hundred females. Hippopotamus societies are matriarchal. Only if the bulls obey a strict code will the females tolerate their presence. A male may approach a group of females only if he is formally invited, i.e., only if a female calls to him. The male must assume a humble posture. If even one female in the group rises to her feet, the bull must immediately lie down. He may stand up only when all the females have lain down again. If a male commits the slightest faux pas, all the females in the area will join forces and drive him away.

young birds become completely independent and leave their families behind.

When the young birds have left, the parents have no reason to stay together. Male and female Tasmanian devils separate when their offspring are gone, as do male and female penguins and seagulls. Moreover, at this time of year, the birds no longer feel any sexual drive.

On the other hand, for the time being there is nothing to prevent the male and female shrike from staying together. The food supply is abundant until the end of September. Since both birds can easily find enough to eat, neither will begrudge the other its food. And since there is no pressing reason for the mates to separate, the personal bond between them holds them together.

Sometimes the shrike mates travel south in August. If they remain in the north, they separate in autumn, when food is in short supply, and occupy separate winter territories. At this time of year the shrikes are so quarrelsome that a male and female will battle to the death if they are put in the same cage.

Despite their marital problems, the male and female shrike do not separate altogether. They live in neighboring territories and maintain long-distance contact. From time to time both birds utter long-drawn-out cries sounding like "kuweet." Then they streak high into the sky and fly long distances over each other's territories, as if each wanted to assure the other of its presence and check up on its mate's whereabouts.

At the beginning of March, just before the next mating season, the male and female shrike are reunited. They do not engage in an elaborate courtship as shrikes do when they do not know each other. The mates already know each other well, and as soon as the food supply permits, they live together as they did the previous year. Thus woodchat shrikes live in intermittent but permanent monogamy.

# CHAPTER 19

## Is Fidelity an Illusion?

*Enduring Monogamy*

Skeptics may claim that no marriage lasts forever, but they are wrong. Ideal marriages do exist. In these marriages, the partners are always faithful to each other, are never apart, do not quarrel, and never get a divorce. In fact, the two mates are indissolubly bound together forever. They grow into a single entity.

To see such marriages, we must plunge into the eternal darkness of the ocean depths where male and female, mother and child, friend and foe recognize each other not by their shapes but by the patterns of the lanterns they carry, the fluorescent colors they wear, and the rhythmic beat of their blinking lights. This lantern-lit world resembles the world of fireflies. Among its inhabitants are various species of voracious predators belonging to the suborder of deep-sea angler fish, which use their lanterns to entice prey into their waiting jaws. No males and females could be closer than the males and females of these species.

The anterior "spine" of the female's dorsal fin is very long. She can use it as a fishing pole, swinging it forward so that it hangs down in front of her needle-sharp teeth. At the end of the "pole" dangles a small brightly lit ball resembling a glowworm, which serves as the fish's bait.

In the dark water, the deep-sea angler can see glowing animals such as brightly lit fish, cuttlefish, and small crabs.

Beside the males, the female deep-sea angler fish is a terrifying giantess. The mouths of two tiny males have become inseparably joined to her body.

However, some animals are invisible to the eye. These are creatures that do not glow in the dark or that darken their lanterns when attacking prey. The deep-sea angler fish knows when such invisible creatures are approaching, for organs along the sides of her body register the vibrations emanating from nearby objects and pinpoint their location. The moment the prey lunges forward to take the bait, the deep-sea angler fish retracts the bait and springs forward to attack.

Johnson's black angler fish can swallow prey twice the length of her own body. Deep-sea angler fish resemble some people in that they have a big mouth with nothing much behind it. The fish have tiny stomachs. However, their stomachs behave like balloons that can be expanded to four times their normal size. Thus a deep-sea angler fish can swallow a large lantern fish, tail and all.

All deep-sea angler fish have dark coloring that absorbs light. Many deep-sea fish and cuttlefish have such bright "spotlights" that if the angler did not wear protective coloring, they could light up her shape and thus see through the trick with the fishing pole.

Their dark coloring poses a problem for deep-sea angler fish. The only light they possess serves as fishing bait, and the rest of their bodies is invisible in the dark waters. How then can male and female find each other during mating season? We have already noted that the difficulties of finding a mate led butterfly fishes to practice lifelong monogamy, despite the fact that they do not have to remain together to care for their young. Deep-sea angler fish face the same problem, and they have solved it by a method that is virtually unique in the animal kingdom.

Apparently a scent may attract a male angler to a female. When he finally locates a female in the vast ocean depths, the male bites into her body and holds on tight. Fearing that he may lose her again or that she may devour him, he never looses his grip. From now on the two of them are literally inseparable.

Gradually the area around the male's mouth becomes part of the female's body, and the "love bite" becomes a permanent kiss. Originally the male's eyes were very large, but once he becomes attached to a female, they degenerate and he goes blind. His mouth and digestive organs atrophy. The circulatory system of the male merges with that of the female. From now on, he receives his food through the female's blood, being nourished as unborn children are nourished through the placenta. However, unlike an unborn child, the male fish does not live inside the uterus, but remains attached to the outside of the female's body.

In short, the male becomes a parasite living on the female. His sole task of life is to produce sperm to fertilize the female's eggs. With her "built-in" male parasite, the female almost constitutes a hermaphrodite. This most devoted of the world's male animals is a highly specialized creature designed purely to fulfill his sexual function. Like the male green bonellia, he is physically tiny compared to his wife. Once a female Greenland angler, some forty-six inches in length, was fished up from the depths of the North Atlantic. None of the three males fastened to her body was more than half an inch long. The female weighed almost a million times as much as each of her consorts. In human dimensions, this would mean that a woman would carry around a husband the size of a small birthmark.

After some time, the sexual organs of the male angler fish grow until he reaches a length of almost four inches. At this point the rest of the male in effect represents nothing more than a container for his sexual organs.

Another primitive method of ensuring the fidelity of a mate is to imprison him or her. The male of many hornbill species walls up his mate in a space about twelve by sixteen inches in the hollow of an old tree trunk. As soon as the female has laid her two to five eggs inside the hole, the male seals the entrance with a mixture of chips of wood, feces, saliva, and mud. He leaves a narrow crack in the wall. The female, which is about the size of a hen turkey, sticks her five-inch-long bill through the crack and jabs at enemies like apes or snakes. She also eats food that the male brings her.

Thus hornbills are just the opposite of deep-sea angler fish: The females become the parasites of the males. The females have voracious appetites. During each visit, the male must fetch some sixty pieces of fruit to feed his mate. In a single breeding season, scientists observed a male bring twenty-four thousand pieces of fruit to his mate. She ate an entire fruit market!

When scientists began to investigate the food a male hornbill brings the female, they at first mistakenly believed that he was trying to poison her. The male fed the female nuts containing a quantity of strychnine that would be fatal to a human being. However, it turned out that the female was immune to the poison.

But what becomes of the imprisoned female hornbill if the male is killed? Is she doomed to starve to death, or can she manage to free herself? She could in fact break out of her prison, but it would do her no good to do so. As soon as the male walls her up, the female begins to molt and loses almost all her feathers, which she uses to line the nest for herself and her young. Now that she is almost naked, she can no longer fly. Thus inside her prison or outside, the widow is doomed unless another male appears to take care of her.

During mating season, young bachelor hornbills patrol the nest holes of their neighbors, watching to see whether males with mates are doing their conjugal duty: feeding the females. As long as a male brings food to the female every day, his nest and his mates are off limits to bachelor hornbills. However, if a bachelor observes that a female is being neglected, he will conclude that she is a widow and begin to feed her as zealously as her own mate. From then on, despite the fact that they have never had sexual contact, the two birds will be mates. Thus the bonding instinct alone, independently of the sexual drive, can be strong enough to establish a relationship between two animals.

The two hornbill mates remain together even when the young birds are fledged, the prison wall has been broken down, and the whole family has begun to fly around outside. By this time the female has grown a full coat of feathers. After the young birds fly away, the adults live together for the rest of their lives. Thus in reality, no wall is needed to keep two hornbill mates together. If the female were not a prisoner, she would still be faithful to the male. The wall merely serves to protect the female while she is molting and hatching her eggs.

Some black African tribes, misinterpreting the behavior of

Two huia mates.

the male hornbill, used to wall up their wives inside mud huts. They kept the women prisoner not just during pregnancy, but all their lives.

Sometimes one animal partner literally cannot live without the other. This was true of the now extinct huia, a species of bird native to New Zealand.

Huias used to eat insect larvae that bored tunnels deep inside the wood of trees. The male huia could not reach the larvae without the help of the female, and vice versa. Huias were large black crowlike birds about eighteen inches long. The males had strong short beaks that they used to strip the bark from tree trunks. However, their beaks could not reach into the narrow tunnels where the larvae were hiding. The females had thin, saberlike beaks that curved downward. These beaks were ideal for digging out the larvae, but were too weak for prying off.the bark. After the male had removed the bark, the female picked out the larvae and fed some of them to her mate. Neither bird could survive without the other, and thus the two mates were forced to be faithful to each other.

Young huias had to mate at an age when they were still being fed by their parents, for no bachelor or spinster bird could survive alone. Widows and widowers were also doomed to starve. As a result, huias were probably the most faithful mates in the entire animal kingdom.

Unfortunately, we will never learn further details about the married life of the huias. In 1907, Maori hunters shot the last of these fascinating birds. Huias were easy prey, for the hunters knew how to imitate their distress call and used the call to lure the birds to their death.

We can only guess how the huias employed their distress call. If one mate died, the surviving mate must soon have grown hungry and called for help. Probably neighboring couples answered the call, feeding the widow or widower until another bereaved partner or a young bird without a mate could fill the place of the mate that had died.

Few species besides the huia have realized the ideal of life-long marital devotion. To be sure, many species practice life-long monogamy. Among them are certain isopods, some species of crab, cichlid fish, butterfly fish, countless varieties of songbird, ravens, doves, greylag geese, parrots, beavers, jackals, dwarf antelopes, some whales, badgers, marmosets, and gibbons. However, the married lives of these creatures rarely resemble a human being's conception of what a good marriage should be. Unless monogamy is essential to survival, as it was for huias, animals, like human beings, tend to be unfaithful to their mates.

Lovebirds lead far from exemplary love lives. Yellow-collared lovebirds closely resemble their relatives the budgerigars in size and appearance, except that their heads are covered with black feathers. The lovebirds live in flocks on the grassy plains of the Tanzanian highlands and build their nest holes in the trunks of solitary acacias and baobab trees.

The birds hold amusing contests to determine their social status in the flock. Two males or two females sit on a branch and stare at each other, at the same time rocking back and forth and from side to side. Suddenly one bird lunges forward, pecks the other bird's foot, and raises the foot in its beak, causing the other bird to loss its balance and topple from the branch.

The turbulent, squawking flock of lovebirds consists of married couples and "teen-agers" looking for mates. Evita was a teen-aged lovebird that had just reached sexual maturity.[1] A high-ranking female, she was courted by no fewer than five males at a time. None of the males wanted to let the other males near her. Soon the two lowest-ranking suitors gave up and turned their attention to less desirable females. Enzo, the highest-ranking male, Memo, the second in rank, and Peppino continued to vie for Evita's favor.

At first the powerful Enzo appeared to be the victor, for he could drive away the other two males whenever he chose. However, Evita had other ideas. Enzo made charming bows, trotted back and forth in front of her, and gradually tried to come closer; but whenever he approached, Evita drove him away.

It was not clear why Evita rejected Enzo's suit. Did she simply find him unappealing? Did she dislike his cocksure manner? Or did she resent the fact that, at the same time he was courting her, he was also courting three other females? In any case, she was not dazzled by his high rank. After she had snubbed him seven times, Enzo gave up and turned his attention elsewhere.

Soon Memo too abandoned his courtship of Evita. She had chosen Peppino, the lowest in rank of her three suitors. When Evita and Peppino were formally committed to each other, all the other lovebirds, including the higher-ranking and stronger members of the flock, respected their choice.

Turbulent as their courtship may be, once two yellow-collared lovebirds become mates, they remain faithful to each other for the rest of their lives. However, if one lovebird mate should die, the survivor promptly seeks a replacement. He or she will give preference to former suitors or "girl friends" if such are available.

R. A. Stamm [2] conducted a series of experiments to determine how faithful lovebirds were under abnormal conditions.

He kept a flock of birds in a large aviary divided in two by a glass partition. Lovebird mates were placed on opposite sides of the partition. The mates could see each other through the glass, but the presence of many other birds gave them ample opportunity to seek other sexual partners.

Romeo and Juliet were two of the mates separated by the glass partition. For months the two of them crouched next to the glass, calling and looking at each other. When mating season arrived, both of them mated with new partners. However, whenever they had a spare minute they returned to the glass to gaze at each other. Clearly the personal bond between two animals can prove stronger than the sexual drive.

When the glass partition was removed, both Romeo and Juliet instantly abandoned their new mates and once again became inseparable.

Not all the lovebird mates behaved like Romeo and Juliet. Each couple in the series of experiments behaved a little differently from all the others. Some mates soon stopped looking at each other through the glass and devoted themselves exclusively to their new partners. Others behaved as if their new partners were the true "loves of their lives." Thus when the partition was removed, a number of birds did not return to their former mates.

Three factors determined whether a couple would be reunited after the partition was removed. These factors were: (1) the strength of the sympathy bond between the former mates, (2) the strength of the sympathy bond between the new mates, and (3) how long the former mates were separated.

In this book I have made it clear that the sympathy bond between two animals can persist much longer than any sexual tie. However, as lovebirds demonstrate, this bond does not always persist throughout life. If two mates are separated for a prolonged period, they may forget each other. By the same token, if a human couple are separated, they may grow apart. For example, returning prisoners of war often find that they no longer have anything in common with their wives.

Sometimes, when two lovebird mates were separated by the partition, the male would remarry very happily, whereas the female would be unhappy with her new mate and would spend all her time gazing longingly through the glass. When the partition was removed, the unhappy female might try to return to her former mate, only to be rejected.

What happens to monogamous lovebird unions when there are many more females than males, and vice versa? Stamm's experiments indicate that when females greatly outnumber males, the birds change their mating habits. Under normal conditions the males court the females and the females have the freedom to accept or reject them. However, when females outnumber males, the females without mates take the initiative and court the males, including males that already have mates. A jealous wife will try to drive away her rivals, but frequently she is unsuccessful, and her husband may end up with two or three wives.

If a male lovebird has two mates, the females raise their young in separate nests. However, after some time the females will grow to tolerate each other if they meet away from their nests.

The behavior of lovebirds when there is a surplus of females recalls the leviratical marriages of the Old Testament. According to Hebrew law, the brother of a man who died leaving no children was obliged to marry his brother's widow and beget a son to serve as the son and heir of the dead man. He had to marry the widow even if he was already married. In those days, when many men died in battle, many women were left without husbands. Thus the few available men had to "spread themselves

Animals that keep harems differ from monogamous animals in two respects. First, males fight bitterly for the possession of a harem. (The upper picture shows two red deer bucks fighting during mating season.) On the other hand, not physical strength but a personal bond between the partners forms the basis of a monogamous marriage. Second, unlike monogamous males, the pashas, or rulers of a harem, never enjoy a permanent relationship with any of the females in their domain. The bottom picture shows a male fallow deer tenderly licking a female. However, no matter how tenderly a pasha treats a female, the bond between them never persists for more than a couple of days. The more or less permanent harems of Oriental potentates are unique in the world of nature.

around" like male lovebirds when there is a shortage of males.

However, when male lovebirds outnumber the females, the females do not resort to polyandry. Males without mates may be "friends" with a married female and are even allowed to feed her when she is sitting on her eggs; but that is the full extent of their services. However, if a female's mate should die, her "friend" will become her new mate. Thus under laboratory conditions, male lovebirds practice the same "deputy system" as geladas and hamadryas baboons living in the wild.

Like many bird species, lovebirds have developed the technique of feeding their partners into the fine art of billing. Lovebirds pass food to each other on the tips of their tongues so that only the tips of their beaks actually touch. When they feed each other, the two birds cock their heads on opposite sides. Thus their billing resembles a human kiss.

Lovebird mates frequently "kiss" one another. For example, they kiss in greeting whenever they have been separated for even a minute; when they are threatened with some danger like an approaching snake; when they are sitting side by side on a branch; before mating; and whenever they make up after a quarrel.

Only lovebird mates ever kiss each other. Fiancés and "deputy males" are only permitted to feed the females, not to kiss them. It is amazing to observe how precisely animals regulate their behavior so that it is always appropriate to their own social status and the status of other members of their group.

It is the personal bonding instinct, the strong sympathy bond between two lovebirds, that enables lovebird mates to remain faithful to each other all their lives. The stability of an individual lovebird union depends on the strength of the mutual bond between the mates. Moreover, the marital behavior of lovebirds mirrors human behavior. Their example should stimulate us to investigate our own bonding instinct so that we can better understand those forces that exercise such a crucial influence on human marriage.

Mephisto's question in *Faust,* "Now I ask you, tell me if you will, why man and woman get along so ill," is now more relevant than ever before, for never have marriages seemed as unstable as they do today. Marital disharmony and divorce are undermining an institution that for many people serves as their principal source of emotional security.

I will not discuss here the effect that unhappy marriages and divorce can have on children. Suffice it to say that the quality of our marriages shapes the quality of the human beings on whom the future fate of the world depends.

I believe that many modern marriages are unhappy because the two partners mistakenly believe that sexual attraction is the primary component of a happy marriage. A marriage based solely on sexual attraction will be as barren and unstable as a marriage based on the desire for money or social position. Songs, poems, plays, novels, films, and television scripts frequently paint a false picture of what love is really all about. These false images of love fail to show the importance of a close personal bond between the partners as the foundation of a happy marriage.

Scientists have only recently discovered the existence of the social bonding instinct. It seems significant that this drive was discovered not by philosophers or sociologists engaged in studying human beings, but by zoologists studying animal behavior. The bonding instinct shows that the behavior of animals can be far from "animalistic."

I have indicated the strength of the tie that binds two lovebirds together. Next I will discuss the married life of animals that come less close to achieving ideal monogamy.

The bonding instinct functions less efficiently in budgerigar unions than it does in the marriages of their close relatives the lovebirds. Budgerigars mate for life; but both partners, particularly the males, are often unfaithful. However, the male's tendency to stray does not affect the stability of a budgerigar marriage. For these birds, a sexual relationship and the personal relationship between two mates are two separate things. As a rule, a male budgerigar is not the real father of the young birds he raises with paternal care. However, the birds are unconcerned with questions of paternity, for of course they do not understand the relationship between copulation and the begetting of young.

For a budgerigar, social and sexual ties are completely

distinct. As a result, it sometimes happens that two birds of the same sex set up housekeeping together. Two males or two females enter into a permanent union. If both birds are females, they each have liaisons with male birds and then lay their fertilized eggs in the same nest and raise their young together.

Like budgerigars, human beings can sometimes separate their sexual feelings from feelings of affection. I once saw an English comedy film called *The Captain's Paradise,* about a ship's captain who used to travel back and forth between Gibraltar and Tangier and had a wife in each port. In Gibraltar he was married to a domestic, home-loving woman. In Tangier his wife was a sexually stimulating belly-dancer. The captain satisfied his intellectual needs by talking with the famous passengers on board his ship. He had in effect divided his life into three separate compartments. As a result, he had to be constantly on guard for fear that one of his wives might learn of the other's existence. Eventually the inevitable happened and the captain lost his paradise.

In ancient Tibet, it was the custom for the master of the house to allow overnight guests to have sexual intercourse with his wife. His wife's "infidelity" did not damage their marriage, and the husband never questioned the paternity of her children. In other words, he regarded marriage and sexuality as separate issues.

Sometimes a man may yield to a momentary sexual attraction and be unfaithful to his wife. The wife need not always take such an infidelity seriously, for the husband's affair does not necessarily mean that he does not love her. When one partner chooses to divorce the other for infidelity, he or she may be using the infidelity as an excuse to escape from a marriage that has already lost its meaning.

However, the unfaithful person himself should always take his infidelity seriously, for he is playing with fire. Sympathy and sexual attraction go hand in hand, and when one is present, the other may soon develop. Thus when a man is physically unfaithful to his wife, he runs the risk of being emotionally unfaithful too, for he may come to love as well as desire his mistress.

A man may have a number of close male friends, but he

has room in his heart for only one woman. Even men who keep harems always have a favorite wife. Thus the formation of a new sexual bond can destroy an old sympathy bond.

Moreover, as the behavior of bearded tits reveals, an animal that has been subjected to sufficient emotional stress can lose the ability to form a sympathy bond.

Bearded tits inhabit the belt of reeds surrounding Lake Neusiedler on the Austro-Hungarian border. The belt of reeds is more than half a mile wide. Some four to seven young birds occupy each of the nests in the wide-flung colony. In June the young birds are fledged. They spend their nights quietly huddled side by side in the reeds, their feathers fluffed out so that they look like furry balls. But at dawn they begin to utter raucous cries, and the males peck and jab and tug at each other and above all at the females. Although they are not yet sexually mature, the young birds are looking for their future mates. The males behave just like little boys teasing and nagging little girls. At first a male will persecute all the females in sight, but then he begins to concentrate his attentions on one particular female. If she patiently endures all his irritating pranks, the two birds consider themselves engaged. They do not mate until the following spring; but once they are engaged, they remain together for the rest of their lives. However, if they find that they are not as compatible as it first appeared, they may in very rare cases break their engagement.

Animals may enter into an engagement, a sort of "trial marriage," to test their mutual compatibility. They do so before reaching sexual maturity. Thus the engagement is a response not to sexual feelings but to the pair bonding drive based on personal sympathy. If the bond between the partners fails to stand the test of time, they can break the engagement.

If an engaged pair of bearded tits get along well together, the male will not stray from the female's side. Otto Koenig [3] states that when the two birds are bathing, preening, sleeping, or looking for food, they always stay close together. The pair caress each other by running their beaks through each other's feathers. If one bird flies on to the next tall reed, the other immediately follows.

Once two birds are engaged, the rowdy young teen-agers

that have not yet found mates will leave them alone. If another young bird dares to attack one of the engaged couple, the pair will give him a sound thrashing. Thus one advantage of being engaged is that both birds lead more peaceful lives than they led before the engagement.

As the weeks pass and more and more birds became engaged, the males that have not yet found a partner become frantic and begin to proclaim their despair to all the world. The song they sing sounds like "chin-jick-chhray." This is a composite song made up of several sounds whose meaning scientists have been able to decipher.

Otto Koenig discovered that in the language of the bearded tit, "chin" means something like "Pay attention!" or "Look out!" "Jick" means that the male bird is in a romantic mood. "Chhr" is a call to the female, and "ay" is a plea for sympathy because the bird is alone. Thus we can translate the male tit's song to mean: "Pay attention! I'm in the mood for love. Female, come here! I'm so lonely!" Young engaged males call "jick-chrr," or "I'm in the mood for love. Female, come here!" In other words, they do not call upon all the members of the flock to pay attention, and they do not express their loneliness.

If, while busily looking for food, a female bearded tit loses sight of her fiancé, she will call "chhr-chhr." Immediately the male answers "jick" and flies to her side. Only her fiancé will answer the female's distress call; the other males will ignore it. Thus her call represents more than a passing expression of the female's mood. Instead it is a consciously directed, purposeful signal, a personal summons sent to a particular bird. In the fifties, scientists investigating language did not consider animals capable of exchanging purposeful signals. The purposeful exchange of signals may be one means of instituting or maintaining a sympathy bond between two animals. It may also help to soothe their aggression.

Desmond Morris [4] pointed out that many people introduce essentially meaningless or irrelevant topics into their conversation. This apparently purposeless chatter serves to maintain contact between two people, much as the practice of grooming maintains contact between two apes. It has no real function outside the immediate context of the conversation.

Bearded tit mates love each other so tenderly that at night the husband spreads out his wings to cover his wife and keep her warm.

Bearded tit mates can be separated only if they are locked in separate cages. If one mate is killed by a hawk or a cat, the survivor flies around calling out to its partner. Then it sits quietly, apparently overcome by grief. Otto Koenig points out that if the bereaved bird hears a rustling in the reeds or the cry of another tit, it will become very excited, as if it momentarily expected its mate to fly to its side.[5]

After a prolonged period of mourning, the surviving mate may marry again. At this point its range of choice is limited, for the nerves of a sober two- or three-year-old tit cannot tolerate the quarrelsome, noisy behavior of the younger birds. The widow or widower must join the flock the adult birds form when it is not mating season. The adult birds are quiet and peaceable. They do not court each other with passion and are never jealous. If a widow and widower meet under these conditions, they may enter into a *"mariage de convenance."*

As the behavior of bearded tits reveals, once an animal has formed a close bond with a mate, he or she may be emotionally exhausted and unable to form such a bond with a second mate. A widowed bearded tit can easily produce young with its new mate. However, frequently the two birds quarrel a great deal. Moreover, they spend more time apart than tits involved in their first marriage. Sometimes they separate altogether or "get a divorce." Thus older bearded tits engage in a staid and sober courtship, and as a rule they find that a second, "sensible" marriage is not as happy or as durable as one forged in the passion of youth.

Of course, there are exceptions to this rule. Occasionally a bearded tit may be unhappy in its first marriage. If its first partner dies and it mates with a compatible partner, it may be far more contented than it was before. However, in general birds which have already lost one partner engage in an apathetic courtship and fail to form strong ties with their new mates.

A natural law determines the strength of the marital bond between animals or human beings. The strength of the bond depends on the strength of the bonding instinct in both partners—i.e., on the innate capacity of both partners to form a strong attachment; on how compatible the two mates are; and on the degree to which each mate has already exhausted his or her potential to form an attachment.

# CHAPTER 20

# From Animal to Man

*Monogamy as a Cultural Creation*

Most of the monogamous animals I have discussed have been birds. In their marital relations, these bird species display many of the all-too-human failings of our own species. However, our closest relatives in the animal kingdom are not birds but apes. Are the mating habits of apes similar to those of human beings?

Most of the monkeys and anthropoid apes are not monogamous. We have already noted that anubis baboons live in large troops and practice free love. Male geladas and hamadryas baboons keep harems and beat the females to keep them from running away. Thus none of these species practices a form of marriage related to monogamy in human beings.

Some troops of Japanese macaques practice free love like anubis baboons. Other troops obey a patriarchal system in which the males treat the females like slaves. Still other troops are matriarchal. (I have discussed these diverse social systems in my book *The Friendly Beast.*[1]) Each troop develops its own tradition, its own form of sexual and social behavior. However, Japanese macaques never practice monogamy.

The only monogamous apes are the gibbons and marmosets that inhabit the jungles of South America and southeast Asia.

The gibbons and their close relatives the siamangs are the aerial acrobats of the tropical rain forest. They belong to the superfamily of anthropoid apes and are also known as "Old World howler monkeys." Their song carries for miles and is weirdly beautiful. Travellers have described the strange concerts which siamangs perform at dawn in the mountainous jungles of Sumatra. As soon as the sun drives away the morning mists, the apes begin their uncanny chorale. Several females utter a series of deep, bell-like tones that resound among the trailing vines. Then the males, which are about three feet in length, give a shattering cry followed by a shout of jubilation and then by a diabolical titter. Finally the males burst into a yodelling song of joy. Group after group of siamangs joins in the song until the jungle resounds. Then after around twenty-five minutes, the primal music slowly grows softer and dies away.

The "singing contest" of the siamangs is a highly disciplined performance. In 1970, Jürg Lamprecht [2] established that the monogamous siamangs follow precise rules in singing their duets. The male and female sing particular sounds in a particular order, and each knows exactly when to chime in and join its partner.

The musical performance of the siamangs resembles the behavior of an opera company during rehearsal. After a brief introductory howl, each ape primadonna sings a series of long bell-like tones that come shorter and faster until they turn into a hellish laughter. Meanwhile the tenor sits indifferently in the wings, ignoring the vocal exercises of his mate. However, the louder and livelier her staccato notes become, the more attentive the male appears. Without making a sound, he rhythmically opens and closes his lips as if he were getting ready to join in at just the right moment. Finally, as his mate barks faster and faster, he is carried away by her virtuosity. Rising abruptly, he utters a prolonged cry followed by a whoop of joy. The moment the whoop comes to an end, the female ceases her diabolical laughter. The first stanza of the duet is over.

Now the male sits down, usually with his back turned to the female. After a pause of several seconds, the female begins the second stanza just as she began the first. As the tempo of her laughter increases, the female swings faster and faster through the trees. Meanwhile, after the female has uttered only two or three cries, the male chimes in with short howls followed by a long drawn-out whoop. Racing around madly, he challenges his mate, attempting to bark even faster than she does. Sometimes one of the apes rhythmically beats its hand against its open mouth like an American Indian letting out a war whoop.

"Slowly the two mates cease their gymnastics," writes Jürg Lamprecht. "Sitting down or hanging quietly from a branch, they utter occasional barks. They do not appear in the least tired out by all their exertion. For the time being they remain silent, moving little or not at all until it is time for their next duet." [3]

Each duet consists of two stanzas and lasts around fifty seconds. The pause between duets lasts for between twenty and sixty seconds. The two mates synchronize their parts during each performance. The joint concerts last for twenty or twenty-five minutes. However, the performers do not always know their parts. If a male siamang fails to join in at the proper place, the female breaks off the stanza, apparently feeling that there is no point in her going on. Occasionally the female may omit her rapid staccato barking in the second stanza. In this case, the male also omits a portion of the duet and synchronizes his music with that of the female.

What the satiric poet Wilhelm Busch wrote about human beings also applies to apes: "In a duet one cannot hide two mouths, both of them opened wide." While they are singing, siamangs blow up their throat sacs like balloons. These sacs, which are larger than a human head, amplify the sound of the music. As a result, the concerts are so deafening that human beings in the area cannot even hear each other speak.

All siamang duets follow the same basic pattern. However, there is always room for individual improvisation. Each siamang couple vary the tempo of their barking and laughter, as well as the pitch of the notes and the length of the intervals between stanzas and duets. Probably these differences in singing technique enable various jungle couples to recognize each other from a distance.

The principle purpose of these duets is to cement the bond between two mates even when it is not mating season. When human beings sing together, they experience a feeling of camaraderie. Probably siamangs feel much the same emotion.

Siamang mates that live in zoos and are unhappily married never sing duets. In the Rotterdam zoo, the female siamang is afraid of her mate, and in the Berlin zoo, the male lives with two females that are too young for him to mate with. In neither zoo do the siamangs sing together.

However, the two siamang couples in the Frankfurt zoo sing quartets. Jürg Lamprecht recorded music showing that each male followed his mate's cues and chimed in at just the right place. Moreover, both the females were careful to begin their stanzas at the same time. The result was a quartet.

In the jungle, neighboring siamang families meet together to sing their chorales. Presumably they are waging a war with music, using song to mark the borders of their territories. Their wild dancing during the song serves to demonstrate their physical strength to other siamangs. After the performance, all the groups leave the border in silence.

The more siamangs inhabit a given region, the more often they sing. In heavily populated areas, siamangs sing not only at dawn and in the evening, but at intervals throughout the day. As soon as one family beings to sing, all their neighbors join in. Soon the entire jungle resounds with baying, laughter, and whoops of joy.

In the Frankfurt zoo, there are no other siamangs to inspire

King penguins have amiable dispositions and rarely quarrel. Because they all get along so well, it is unimportant for a king penguin to return to the mate it had in the previous breeding season. Any penguin can easily take a new partner.

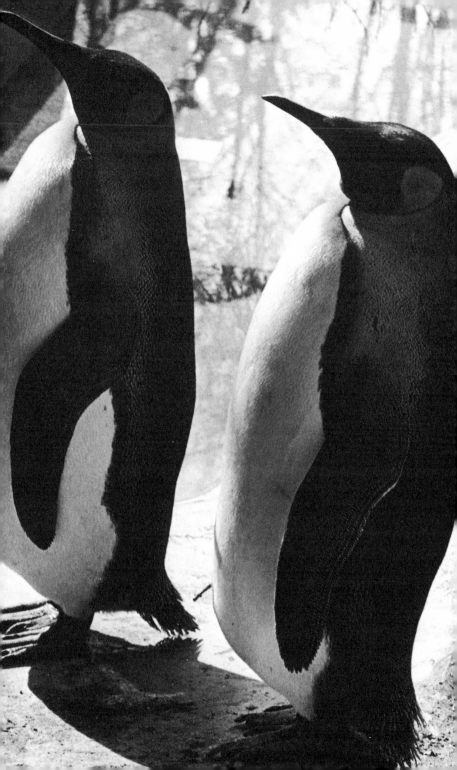

the two siamang couples to sing. Instead they respond to the screaming of schoolchildren who come to visit the zoo.

The members of a siamang family spend the night together in the center of their jungle territory, sleeping in the branches of a tree. For example, in one family the father and mother will curl up close together in the middle of the group, warming a three-week-old baby between them. Nearby are three of their young, which are two, four, and six years old. The twenty-year-old grandfather huddles close to the group. Off to one side sits a young eight-year-old mother with her child.

Male and female siamangs and gibbons are monogamous and mate for life. Both are extremely aggressive toward other members of their own sex. In other words, these apes behave very much like blind goby fish and green lizards: The males are hostile to males and the females to females, but animals of the opposite sex never attack each other. Of course, siamangs and gibbons do not "marry a place" like blind gobies and green lizards. They enter into permanent monogamous relationships and share a strong personal bond with their partners.

A pair of storks mating. If the female is removed from the nest and replaced with another female, the male is perfectly content to mate with the replacement. Storks practice a special type of marriage. They are married not to each other but to the nest.

Two stag beetles fighting. Just as the European giant deer developed giant antlers and the saber-toothed tiger gigantic fangs, the parts of an insect's body may develop monstrous proportions. The "antlers" of the stag beetles are really overgrown mandibles. There are nine hundred species of stag beetles. In some of these species the jaws alone are as long as the rest of the beetle and as broad as they are long: around two inches. No one knows the purpose of these "superweapons." In combat, small agile "Davids" often defeat the clumsy "Goliaths." The females have much smaller mandibles. It is unimportant whether they find the males' mandibles attractive, for female stag beetles are not permitted to choose their mates. Thus there can be only one possible explanation for the size of the male's "antlers": the larger his jaws are, the larger the female he can steal from a rival and carry off into the bushes between these jaws.

There are two exceptions to the rule that gibbons and sia-mangs are always hostile toward members of the same sex. A mother will allow a daughter to remain with the family group even after the daughter has reached sexual maturity and has had her first young. Apparently the mother does not resent the daughter even though the daughter's child must have been begotten by her father—i.e., by the mother's own husband.

While they are still living with their parents, young females bear and raise their first baby as a sort of "practice run." Usu-ally the baby dies. Female apes have to learn how to care for their young by watching their mothers take care of younger brothers and sisters and by practicing on their own offspring.

Gibbons are monogamous. However, before a female finds her partner, she first has a sexual relationship with her father and bears him a child. The father is unfaithful to his mate when he has sexual relations with his daughter.

There is a second exception to the rule that siamangs and gibbons are hostile to members of the same sex. The family does not drive a grandfather away. The grandfather is an aging male that can no longer fulfill his responsibilities as head of the family. One of his sons gradually assumes all his functions. The son that takes over the duties of an aging father resembles the "deputy pasha" that replaces the pasha of a hamadryas baboon harem. However, in this case the social unit is not a harem but a monogamous family unit. Grandmothers are also tolerated in the family group.

In other words, as a rule a gibbon family engages in one long marriage that endlessly renews itself. The old male is suc-ceeded by a son that marries the middle-aged wife of the father. Several years later, the female becomes very old, and a daughter takes her place, marrying the son. Thus the father always mates with one of his daughters and the mother with her son. For untold generations, gibbons have practiced inbreeding.

Sometimes these anthropoid apes break the cycle of in-breeding. When they hear the morning chorale begin, the entire family races from the center of their territory, where they have spent the night, toward the border where they are to sing. As they move through the trees, they hear their neighbor's territory resounding with warlike cries. Then the troop speeds its tempo, racing through the jungle.

No one has yet been able to measure how fast gibbons can travel, but there is no doubt that they are the fastest animals in the world. Tarzan swinging from vine to vine seems a snail in comparison. Gibbons gain impetus from using springy branches much the way pole-vaulters use their poles. These apes, which are between eighteen and twenty-five inches in length, travel about forty feet at a bound. They move so fast that they can even capture birds in flight. If a gibbon misjudges its distance and misses a branch it was aiming for, it will simply catch hold of a lower branch with one of its long, muscular arms. Gibbons rarely travel along the ground.

The entire gibbon family race to the border of their territory, where they meet their neighbors and begin a deafening singing contest. Like many bird species, gibbons use song to mark their territories. If one group of gibbons crosses into the territory of another group, the second group will pursue the intruders in fury. However, gibbons rarely engage in physical combat.

The gibbon neighbors work off their aggression by singing at each other for half an hour. Then all the gibbons grow calmer, and neighbors eye each other with hostility through the leaves. At this time, young gibbons from enemy territories are allowed to play together. However, they remain some distance from their parents. Now a young mother that has given birth to her first child may make friends with a young male whose father is always quarrelling with him and will soon drive him away from the troop. The two may become engaged in this no man's land. If they prove compatible, the young gibbons will leave together to find their own territory and raise a family.

Gibbons resemble seagulls in that neither mate begrudges the other its food. Gibbon mates never try to take food away from each other. Often they will share a single piece of fruit. Male and female gibbons behave very much like human beings in love in that they spend long periods embracing and caressing each other. However, in other respects they are quite different from humans. I have already mentioned that, although gibbons are monogamous, their form of monogamy prevents them from forming a social order more complex than the individual family unit. As the young grow older, they are driven away from the troop, and neighbors are regarded as rivals.

The structure of gibbon society makes it impossible for gibbon mates to be unfaithful to each other except with other members of the family group. Neither partner is ever unsupervised for a moment; thus neither has the freedom to make contact with gibbons in a neighboring group. If human society resembled that of gibbons, this would mean that a man would always be by his wife's side and would never let her speak to a male neighbor, salesclerk, bus driver or repairman, or any other man whatsoever. By the same token, the wife would never let her husband approach within thirty yards of any other woman. This behavior would make it impossible for human beings to develop a social unit more complex than the family. Fortunately, we are not as strictly monogamous as gibbons. If we were, human society would consist of nothing but isolated family units, all hostile to one another.

If human beings lived like gibbons, we could not even join together in groups to go hunting. In other words, our ancestors could never have left the jungle, which provided them with food and protected them from the predators of the savannas and plains. All apes that venture onto the plains travel in large groups. These apes include anubis baboons, Rhesus macaques, Japanese macaques, and chimpanzees. It is no accident that none of these species is monogamous. By the same token, it is no accident that the monogamous gibbons and siamangs remain in the jungle. (See the discussion of turkeys in Chapter 15, in which we noted that turkeys that inhabit the forest live in small groups, whereas on the plains they live in large flocks.)

Now let us examine the sexual relations of the chimpanzee, our closest relative in the animal kingdom. The ancestors of the chimpanzee lived in the tropical forest. Forming large groups, they ventured onto the plains. For hundreds of thousands of years, their descendants, the present-day chimpanzees, have lived on the plains and the savanna. However, the ancestors of man and man himself have persecuted the chimpanzee so that in many areas, chimpanzees have returned to their ancestral jungle.

Various factors influence the social behavior of the chimpanzee. For one thing, these anthropoid apes remain children for a long time. Thus a mother must care for her young for seven years. Only human beings care for their offspring longer than seven years. In many nations, laws demand that human parents be responsible for their children for twenty-one years.

Due to the difficulties of finding food and avoiding predators, the female chimpanzee can care for only one young chimpanzee at a time. Thus a female mates only once every seven years. She is sexually mature and able to mate for no more than a few days.

A chimpanzee mating is a brutal and shocking act. All the males of the troop line up and rape the female one after another. The female screams, shows all the signs of extreme fear, and repeatedly attempts to run away and hide—but in vain. On the following days she is raped again and again. When the ordeal is over, her sexual life is over for another seven years. Clearly monogamy cannot develop under such conditions.

Moreover, the bonding instinct is not strongly developed in chimpanzees. Chimpanzees enjoy the company of their own kind and have personal relationships with other chimpanzees, but they do not form lasting ties. When two troops of chimpanzees meet, the members of both groups scream a great deal, show off their strength, and behave somewhat aggressively. However, soon the two troops make friends, and all the chimpanzees utter exclamations of joy and begin to embrace each other. When the groups separate a few hours or a few days later, a number of chimpanzees will change troops.

Chimpanzees are individualists. Thus for both sexual and social reasons, they do not enter into the bonds of matrimony.

Some people mistakenly believe that the higher an animal's place in the scale of evolution, and the more intelligent the animal is, the more admirable its behavior must be. These people will be disappointed to learn that the sexual behavior of our closest relative the chimpanzee is far from edifying. Clearly natural law does not observe the prejudices of human morality.

The fact that a species has evolved quite recently does not imply that it is "superior" to its predecessors.

Species that are very similar tend to compete, so that only one of the species survives. Thus species tend to evolve toward increasing dissimilarity. Therefore, rather than asking why the marital relations of chimpanzees and human beings are not more alike, we ought rather to ask why they are so different.

In human beings, marriage can take many forms. Among the prerequisites of human forms of marriage are intelligence, the ability to transmit learned experience to other humans, the ability to use weapons and tools, and the ability to convert caves into dwellings. In prehistoric times, all these factors lessened the risk that human children would starve or be killed by predators. As a result, unlike the female chimpanzee, a woman could raise several children at a time and could remain sexually active throughout this time.

If early humans had raised their offspring one at a time as chimpanzees do, they would have become extinct, for each woman would have lived long enough to give birth to only two children. Given the high infant mortality rate among primitive peoples, human couples could not have succeeded in reproducing themselves.

As human beings became more intelligent, children took a longer time to develop. In turn, this extended period of development enabled them to grow more intelligent so that the species as a whole increased in intelligence.

Human beings were less aggressive than gibbons and thus were able to form social groups that transcended the family. Moreover, their instinct to form personal bonds was stronger than that of chimpanzees.

This fortunate constellation of intelligence and behavioral patterns made man what he is and at the same time left him the freedom to develop a variety of cultural patterns. Thus human beings are capable not only of monogamous relationships (albeit attended by infidelity, quarrelling, and divorce), but also of keeping harems, engaging in other forms of polygamy and polyandry, and establishing patriarchal or matriarchal societies.

All these possible forms of marriage are expressions of our human nature. It is not heredity alone, but intellect and the power of choice that determine what form of marriage is practiced by a particular society or a particular human being. Our remote ancestors were not monogamous. Thus it is an amazing cultural achievement that in many modern societies, monogamy is the predominant form of marriage.

Human beings are unique in the animal kingdom in that unlike gibbons, they are capable of engaging in monogamous relationships as well as of banding together to form more complex social groups ranging from the tribe to the modern nation.

# CHAPTER 21

# Can Brothers Marry Sisters?

*How Animals Learn to Recognize Appropriate Sexual Partners*

When young bullfinches reach the age of six or seven weeks, they begin to flirt with each other. The four or five brothers and sisters in each family have not yet outgrown their first shaggy coat of light gray and light brown feathers, and they are all still under their parents' care. Moreover, they will not be able to breed until the following spring. Nevertheless, they are already busy courting each other. Males and females look exactly alike; thus often brothers end up courting brothers and sisters sisters. Moreover, males and females all behave alike. That is, they all behave like females.

A pair of birds will kiss each other's beaks and run their beaks through each other's feathers. Then one bird will bow down like a female and invite its partner to mount it and mate with it. Instead of mounting the first bird and pretending to copulate, the second bird hops in front of it, bends down, and invites the first bird to mount it. This exchange of invitations can continue for some time.

Even if both young birds are males, they behave as if they were females. No matter what their sex, their relationship must remain purely platonic, for all the birds are sexually immature. Nevertheless, the young bullfinch couples are inseparable. If one bird flies away, the other follows it. The two birds eat and sleep together and protect each other from attack. At the age of three months, the young males begin to feed their partners as males feed the females during courtship. At this point the males begin to behave like males rather than females.

However, the young bullfinch couple are not really mates. they behave like an engaged couple, but they never actually marry. In other bird species, an engagement gives the two partners time to test their compatibility. The six- or seven-month engagement of two bullfinches serves to train them how to behave toward their mates when they become adults.

Near the end of the calendar year, the two engaged birds begin to quarrel more and more frequently. The first broken engagements are those between two brothers or two sisters. Later, birds of opposite sex also go their separate ways.

Gradually the young birds acquire their adult coat of feathers. Flying around not too far from their nests, they look for new partners. Now that they have learned to distinguish between males and females, they look for a bird of the opposite sex. If a female accidentally tries to court another female, she will be attacked and driven away. Moreover, brothers and sisters that may formerly have been engaged behave very aggressively whenever they meet.

The instinctive hostility of older bullfinches toward members of the same sex and toward members of their own families prevents them from selecting inappropriate partners for their second engagement. However, before two birds marry, they must again become engaged. The engaged pair will not be sexually mature until spring, which is still three or four months away.

During their first engagement to a brother or sister, bullfinches practice being married. During their second engagement, two birds test their compatibility. The second engagement begins with a test of sympathy and aggression.

Not the male but the female does the active courting. She flies around looking for a male. When she finds him, she lands in front of him, stretches out her wide-open beak and shrieks at him in rage: "hu-ay." Frightened, the male flies away. The female furiously pursues him until he is totally exhausted and sits quietly on a branch.

If the male bullfinch wants nothing to do with the female, he has no choice but to keep flying until she gives up the chase. If he finds her an acceptable mate, he begins to resist her attacks, and her fury abates. Finally the male gathers courage, jumps at the female, and swiftly kisses her beak. Immediately he flies a safe distance away. But soon the female ceases to attack him and the two begin to bill. The male feeds the female, and both birds caress each other by running their beaks through each other's feathers. Now the two birds are engaged.

In January and February one sometimes sees young bullfinches that are engaged to two birds at the same time. At first they spend most of their time with their first love, their brother or sister, leaving him or her for only a few minutes to look for another partner. However, the closer the bird comes to sexual maturity, the stronger is the aversion it feels for its brother or sister, and the more time it spends with its new partner.

During their first engagement, both bullfinches behave like females. At the beginning of the second engagement, both birds, and especially the females, behave like males. Thus young bullfinches pass through various stages of development. As their behavior alters, so does their "partner schema," or their image of an appropriate partner. A very young bullfinch is attracted only to its brothers and sisters and feels shy and uncomfortable around other members of its species. As it grows older, its tastes change. Suddenly it begins to dislike the company of its brothers and sisters and prefers to spend all its time with strangers. This instinctive change in taste prevents bullfinches from inbreeding. Of course, the birds never become consciously aware that they must not mate with members of their own families.

Under normal conditions, human beings, like bullfinches, do not need to be formally taught to avoid sexual relations with their brothers and sisters. Research conducted in Israeli kibbutzim revealed that human children instinctively observe the incest taboo.[1] Scientists conducted an investigation to determine whether human beings who had grown up in close proximity married when they became adults. They studied a group of 5,538 people, or 2,769 married couples who had grown up in kibbutzim and later married.

Instead of being raised by their parents, babies and children in a kibbutz are raised in small groups by specially trained personnel. All the children in a group are of approximately the same age. A "group mother" cares for all the children in each group. The children see their own parents for only an hour or two each day, after the day's work is done. In most kibbutzim, the children spend the night with the other children in their group rather than with their parents.

Gradually the children begin to feel as if their "group mother" were their own real mother. The family unit here differs markedly from the traditional nuclear family unit, for one mother cares for seven, eight, or nine boys and girls of the same age. The children are not related by blood, but they grow up sharing all their joys and troubles just like real brothers and sisters. Sometimes the youngsters quarrel, but they always present a united front to the outside world.

Children in a kibbutz are permitted a large measure of personal and sexual freedom. At an early age they begin to play sexual games together, and no one ever interferes with these games. However, at the age of ten the youngsters spontaneously exhibit signs of sexual inhibition and begin to experience feelings of shame and modesty. As I have already mentioned, this fact proves that the sense of shame is not exclusively the product of a repressive upbringing, although such an upbringing can greatly intensify sexual inhibitions.

From the age of ten on, the girls and boys in a kibbutz no longer get along as harmoniously as they did before. They tease and quarrel with each other, refuse to play together, and find each other "dumb." Then, as they near adulthood, their mutual hostility disappears.

Whom do these children marry when they become adults? Scientists studied 2,769 marriages in which both the partners had grown up in a kibbutz. In only thirteen cases had children belonging to the same group married in later life. Two thousand, seven hundred and fifty-six young women and the same number of men had chosen as their partners people who had grown up in different groups. Moreover, people who had grown up in the same group almost never engaged in extramarital sexual relations.

The children raised in kibbutzim were not taught to avoid inbreeding. Moreover, no one in the community would have disapproved of a marriage between a man and woman who had been raised in the same group. Children who had been raised together felt very close, had a great deal in common, liked each other, and did not feel shy around each other. Why then were they not inclined to marry? The twenty-six exceptions, the young people who married members of their own groups, provided the answer.

When the twenty-six young "deviants" were questioned, it turned out that before the age of six, one partner in each marriage had spent some time away from the group to which his or her future spouse belonged. All these partners had temporarily moved to another kibbutz, spent a couple of years abroad with their parents, or undergone a long siege of illness. Fifteen years later when they reached adulthood, they felt no aversion to marrying one of their "brothers" or "sisters." On the contrary, the closeness they felt to members of their own group made such a marriage appear very desirable.

Significantly, if the children were absent from their group for a prolonged period *after* the age of six, they did not later feel sexual attraction to members of their own group. If a child left his group between the ages of five and six, he might marry a member of the group. If he were away from age six to nine, he never married a member of his group, even though he had been away three times as long as the child in the former case.

Thus we must conclude that the period between one and six years is a sensitive period in the life of a child, during which its basic emotional patterns become set for life. Between the ages of one and six, a brother is imprinted with a sexual aversion toward his sister and vice versa. This aversion does not destroy feelings of friendship and affection between brother and sister, but it automatically inhibits feelings of sexual attraction.

Moreover, the behavior of children in Israeli kibbutzim reveals that observance of the incest taboo is not innate. Instead, all children who grow up together are imprinted with this taboo, whether they are blood relations or not.

One might assume that children would be imprinted with the incest taboo at around the age of ten, when they begin to experience feelings of modesty and sexual inhibition. However, in reality the imprinting takes place four years earlier. We know nothing of the forces that promote this imprinting, which produces no external change in the behavior of the child.

Many other factors that shape the personality development of children, as well as of young bullfinches, are equally mysterious. For example, one might assume that all young males who play sexual games with other young males will grow up to be homosexual. But at least for bullfinches, just the opposite is true. Engagements formed between bullfinches of the same sex are less durable than those between members of the opposite sex. As a result, bullfinches develop an aversion to members of their own sex, and this aversion prevents them from entering into homosexual relationships.

Among human beings, sexual roles are more complex. I will discuss human sexuality in greater detail in my future book concerning the development of young and the parent-child relationship.

# CHAPTER 22

# Adultery and Divorce

*The Advantages of Monogamy*

In the Arctic Ocean, halfway between the northern polar ice cap and the Spitsbergen Islands, lies Bjornöya, or Bear Island, a large island with a steep rocky coastline that rises some 1,750 feet above the sea. At some points the cliffs jut out like a balcony over the stormy waves.

Every year in May and June, hundreds of thousands of birds come to breed on these towering walls of rock. Quantities of puffins nest on the southern plateau of the island. Guillemots crowd the projecting ledges of rock and flocks of kittiwakes wheel around the cliffs like flurries of snow. Every knob of rock a hand's breadth in width is the site of a kittiwake nest.

Between 1954 and 1966, an investigation was conducted to determine whether monogamy and lifelong fidelity were beneficial to kittiwakes and their young.[1]

After all I have said regarding animal mating habits, it may appear that monogamy is not very efficient form of animal marriage and that it therefore represents no more than an expensive whim of nature. To be sure, monogamy does spare two aggressive animals the necessity of forming a relationship with a new partner every mating season. But does this increase in comfort justify the vast quantities of time and effort that nature expended to develop the monogamous mating drive?

The existence of forces like the sympathy bond cannot be explained solely on the grounds that they contribute to an animal's comfort. They must also contribute to its survival. While investigating the habits of the kittiwake, John Coulson[2] discovered that monogamous marriage increases the number of offspring and also provides for their physical and emotional well-being, thus helping to ensure their survival.

Physically, kittiwakes closely resemble other species of gulls. In German they are called "three-toed gulls" because, unlike other gulls, they do not possess a fourth toe or claw, which projects from the rear of the foot. Their behavior differs markedly from that of other gulls, for they build their nests on the dizzying heights of steep cliffs rather than in the soft dunes along the shore.

Kittiwakes are almost as aggressive as herring gulls. However, if they were to engage in physical combat, they would fall from the cliffs. Moreover, their nests might easily be destroyed. As a result, the birds rarely fight each other.

Some readers might assume that since kittiwakes can fly, they would be in no danger of falling from the cliffs during fights with other kittiwakes. However, this is not the case. For three hundred days of the year, fog and rainstorms blanket the island. During a storm it is very difficult for the gulls to find their nests. But despite the bad weather, they must go out to catch fish, for if they did not fish, they would starve to death. It is a mystery how they manage to locate the island and their own nests in the dense fog. Even more amazing is the acrobatic skill with which they take off and land in the raging winds. Corpses of dead birds along the rocky shore show that sometimes their fishing expeditions prove fatal. Under these conditions it would be dangerous for two kittiwakes to fight in the air. Thus kittiwakes fight by screaming at each other.

This mild form of combat continues from dawn till dusk, especially during mating season. If a bachelor or spinster finds an unguarded nest, he or she sits down in it and waits to see how the owner will react when it returns. Usually the owner already has a mate and is far from pleased at the invasion. If it threatens the intruder, the strange bird will fly away uttering a long piercing cry.

Thus at frequent intervals, the strength of a kittiwake marriage is put to the test by potential rivals. Moments after the mates have left their nest, a tempter arrives. If the tempter is a male and the female mate is the first to return to the nest, the intruder will try to strike up a relationship with her. Similarly, a female intruder will try to seduce the male. Sometimes the strange bird succeeds in breaking up a marriage and steals the male from the female or the female from the male. Thus although lifelong monogamy represents the norm or the ideal form of kittiwake behavior, many kittiwakes do not realize this ideal.

Over a period of twelve years, John Coulson studied the behavior of a small colony of kittiwakes during mating season. He identified each individual bird and kept records of the gulls' marital relations and of how each couple reared their young. Coulson's studies showed that despite many temptations to adultery, sixty-four per cent of all kittiwakes remained monogamous, faithfully returning to their former mates at the beginning of each mating season. Twelve per cent were forced to seek new mates, for their old mates had died of disease, been killed on the cliffs during a storm, or perished while out fishing. However, twenty-four per cent of the kittiwakes were not widowed, but voluntarily separated from their former mates and entered into a second marriage.

The longer a marriage had lasted, the more likely it was that the two mates woud remain together. After a kittiwake pair had been together for more than a year, they would probably stay married. However, there were some exceptions to this rule. Once a female that had had the same mate for five years returned home from a fishing expedition. Apparently she had been attacked by another gull, for her feathers were tattered and two of them were sticking straight up on her head. As usual, she began to greet her mate by rubbing her beak against his. At first the male seemed willing to respond to her billing. But instead of calming their aggression, the ritual billing became more frantic until the two mates began pecking at each other. Thus under unusual circumstances, a soothing gesture like billing can lead to combat.

The slightly disfigured female did not wait for her husband to attack her. Instead she accepted the fact that she had been driven from her home and never returned. Two days later, the male already had a new mate.

Frequently the physical disfigurement of one partner may destroy the sympathy bond between two animals, resulting in a divorce. We have already noted that if a scientist alters the color of a herring gull's eye ring, he can destroy a herring gull marriage.

The marriages of little ringed plovers are equally fragile. These birds engage in seasonal marriages, but as a rule the same partners mate year after year. One little ringed plover couple had mated for three successive seasons. In the fourth year, the male returned to the breeding grounds missing one leg. The loss of a leg was by no means fatal, for the bird could still move around by hopping on one leg and fluttering his wings. He was also quite capable of mating with the female. But the loss of a leg was an aesthetic flaw, and the female promptly left her unfortunate mate.

Thus even a slight physical flaw may be grounds for divorce in a monogamous animal marriage. However, if a couple have been together for a long time, disease, old age, or sexual impotence are not grounds for divorce.

In the kittiwake colony, many of the males and females were ill, senile, or infertile. One might assume that the sexually active partner would abandon the sick or impotent one and seek satisfaction elsewhere. However, this was not the case.

Bullfinches, Bourke parrots, and violet-eared waxbills, which are all monogamous birds, also divorce their partners only for aesthetic shortcomings, never for impotence.

Once again, these examples prove that the sympathy bond

plays a greater role than sexuality in holding an animal marriage together. The impotence of one partner does not damage the sympathy bond between the mates. However, flaws in the appearance of one partner—flaws like feathers jutting out at the wrong angle or a missing leg—can destroy the sympathy bond.

It may appear disadvantageous to a species that flaws in appearance can serve as grounds for divorce. Clearly a union in which one partner is impotent or senile cannot produce offspring. However, when one partner possesses an external flaw, the two mates are still capable of reproducing. Would it not be more practical if animals divorced impotent mates rather than disfigured ones?

In reality there is sometimes good reason for animal mates to get a divorce if they are no longer pleasing to each other; for if they do not get along well, they may have no offspring.

Approximately one quarter of all the birds in John Coulson's kittiwake colony divorced their mates of the previous year. Coulson discovered that almost none of the birds that looked for new partners had had offspring during the previous mating season. Clearly it was not the lack of offspring that caused the divorce. Instead, certain couples had no young because the parents did not get along well. The disharmony between the mates eventually resulted in their divorce.

Their stormy courtship does not give kittiwake mates time to get to know each other well. Thus they may easily prove incompatible. Species like lovebirds, bearded tits, and bullfinches, on the other hand, thoroughly test their compatibility before mating.

Some twenty-four per cent of the kittiwakes in a given colony have unhappy marriages and produce few offspring during their first breeding season. However, mates that have been together for some time more than make up for the offspring their unhappy peers fail to produce. The longer a monogamous pair have lived together, the earlier they begin to breed each year, the more eggs the female lays, and the more young birds grow to adulthood.

The offspring of a harmonious, durable kittiwake union are physically and emotionally better equipped for survival than those whose parents have been divorced one or more times. Every year that a kittiwake couple stay together, they become more skillful parents. As yet we do not know exactly what skills the parents learn or how they learn them; but clearly the offspring of long-term marriages are superior to those of other unions.

By the same token, the more frequently a kittiwake has changed partners, the less capable it is of successfully fulfilling its duties as a parent. For example, a female that has been divorced twice will be a less adequate mother than a female that is struggling to raise her first brood, or a female that has been divorced only once.

Presumably a divorce has such a frustrating and demoralizing effect on kittiwakes that it takes them a long time to regain their emotional equilibrium. Their young suffer even more, for they often die before reaching adulthood. At the very least, the young birds are inadequately equipped for the struggle to survive. Nevertheless, if a marriage is so unsatisfactory that the mates cannot even produce offspring, it is better that they separate and look for new partners with which they may be able to reproduce.

Thus ideally kittiwakes practice lifelong monogamy, but the mates may separate if they are incompatible. Monogamous marital relations are highly beneficial to the young. Sixty-four per cent of kittiwakes that annually remain with their old partners raise their young so skillfully that they make up for the losses occasioned by unions that produce no young because one of the partners is sick, old, or impotent. This fact demonstrates the utility of a strong sympathy bond that unites two mates for the duration of their lives.

The discovery of the sympathy bonding instinct represents a major achievement of modern science. For the first time, this book has described the important role played by the bonding instinct in the choice of a sexual partner, in pair bonding, and in the conduct of a marriage.

For centuries men have understood the instinctual character of sexuality. Recently Konrad Lorenz demonstrated the instinctual character of aggression. Since his work became known, people have tended to view sexuality as the antithesis of aggression, as a unifying force leading to love. Clearly this is an erroneous view, for sexual attraction cannot guarantee a happy or stable marriage. Sexuality is Janus-faced: It is both a social and an asocial force. In 1971, Hubert Markl pointed this out in his popular work, *Vom Eigennutz des Uneigennützigen*: "Sexuality is a force that can unite members of the same species. However, in the animal kingdom no higher form of social behavior can develop from a sexual relationship. Sexuality may indirectly contribute to the formation of social ties: for example, by inhibiting aggression. However, sexuality is not in itself a socializing force, but rather is a bipolar force that is closely linked to aggression. To be sure, two are more than one; yet they are fewer than three. During copulation, there is no room for third parties. Frequently one or both partners will drive away other members of their species; moreover, third parties often take offense at the sight of two animals copulating. Thus sexuality promotes the development of social bonds, but also inhibits their development."[3]

Sexuality can bring together only two animals at a time. Moreover, it cannot hold two animals, or a man and a woman, together once they have satisfied their sexual drive. Thus complex social units cannot develop from sexual ties.

Until recently, we were unaware of the importance of the social bonding instinct, the feeling of personal attachment, which is the true antithesis of aggression. Now that we know about this instinct, we must reassess the theory and practice of marriage.

What I have written regarding signals that attract and repel, love at first sight, mimicry, the problems male and female experience in getting close, the laws governing sexual harmony, the rules of courtship, the various forms of marriage, and the ways in which animals learn to recognize appropriate sexual partners, opens the door into a strange and unfamiliar universe. Despite its strangeness, in this universe we can learn to understand the forces that govern all our lives. Once we comprehend the forces that rule animal behavior, we will be better able to control our marriages, prevent adultery and divorce, and raise our children in loving, stable homes.

Fortunately, unlike much contemporary scientific research, the study of the bonding instinct cannot be abused for destructive ends. Research into the nature of aggression can give human beings ammunition to excuse or justify aggressive behavior and to manipulate other human beings. However, the investigation of the bonding instinct can only benefit our whole society. The time has come for us to contemplate the forces of creation as well of annihilation.

# Notes

INTRODUCTION: Sympathy, the Antithesis of Aggression

1. Frank A. Beach, "Coital Behaviour in Dogs, III: Effects of Early Isolation on Mating in Males," *Behaviour,* 30 (1968), pp. 218–238.
Frank A. Beach, "Coital Behaviour in Dogs," VIII: "Social Affinity, Dominance, and Sexual Preference in the Bitch," *Behaviour,* 36 (1970), pp. 131–148.
2. Konrad Lorenz, *On Aggression* (New York: Harcourt, Brace & World, 1966; London, Methuen & Co. Ltd., 1966).


CHAPTER 1. VIRGINS BEAR YOUNG

1. A. K. Tarkowski, A. Witkowska, and J. Nowicka in *Nature,* 226 (1970), pp. 162–165.

CHAPTER 3. LOVE AT FIRST SIGHT

1. Manfred Curry, *Bioklimatik; die Steuerung des gesunden und kranken Organismus durch die Atmosphäre* (Riederau Ammersee: American Bioclimatic Research Institute, 1946).
2. Helga Fischer, "Das Triumphgeschrei der Graugans," *Zeitschrift für Tierpsychologie*, 22 (1965), p. 300.
3. *Brockhaus Enzyklopädie in zwanzig Bänden* (Wiesbaden: F. A. Brockhaus, 1969).
4. Heinrich Böll, *Das Brot der frühen Jahre*, (Köln: Kiepenheuer & Witch, 1955).
5. Neal Griffith Smith, "Visual Isolation in Gulls," *Scientific Amercian*, Vol. 217, no. 4 (Oct. 1967), pp. 94–102.
6. Erich Hecker, "Sexuallockstoffe—hochwirksame Parfüms der Schmetterlinge," *Umschau in Wissenschaft und Technik*, 59, pp. 465–467 and 499–502.
7. Irenäus Eibl-Eibesfeldt, *Grundriss der vergleichenden Verhaltensforschung* (München: Piper Verlag, 1967), pp. 410–416.


CHAPTER 4. MISGUIDED BEHAVIOR

1. Nikolaas Tinbergen *et alia*, "Die Balz des Samtfalters," *Zeitschrift für Tierpsychologie*, 5 (1942), pp. 182–226.
2. Nikolaas Tinbergen, *The Study of Instinct* (London: Oxford University Press, 1951).
3. Irenäus Eibl-Eibesfeldt, *Liebe und Hass* (München: Piper, 1970), p. 31.
4. Personal communication to the author.


CHAPTER 5. BRAINWASHING AND THE PERVERSION OF EMOTION

1. Nathaniel Kleitman, *Sleep and Wakefulness* (Chicago: University of Chicago Press, 1964).
2. Erwin Lausch, *Manipulation—Der Griff nach dem Gehirn* (Stuttgart: Deutsche Verlags-Anstalt, 1972), p. 164.
3. Hans Jürgen Eysenck, "Nicht philosophieren, sondern experimentieren—Über die Verhaltenstherapie," *Die Zeit*, No. 39 (Hamburg, 1967), p. 46.
4. José M. R. Delgado, "Die experimentelle Hirnforschung und die Verhaltensweise," *Endeavour*, 69 (1967), pp. 149–154.
5. *Stern*, No. 21 (Hamburg, May 17, 1973).

CHAPTER 7. FROM CANNIBALISM TO LOVE

1. Walter R. Fuchs, *Leben unter fernen Sonnen?* (München: Droemer Knaur, 1973).

2. Manfred Grasshoff, "Die Kreuzspinne—ihr Netzbau und ihre Paarungsbiologie," *Natur und Museum*, 94, Book 8 (1964), pp. 305–314.

3. Richard Gerlach, *Die Geheimnisse der Insekten* (Hamburg: Claassen, 1967), pp. 264–266.

4. June Johns, *The Mating Game* (New York: St. Martins Press, 1971; London: Peter Davies, 1970).

5. Maurice Burton in "Animal Life," *Purnell's Encyclopedia* (London, 1969), pp. 1656–1658.

6. The author, accompanied by a film team, experimented on these fish. No written publication.


CHAPTER 8. AM I MALE OR FEMALE?

1. Robert Ardrey, *Der Gesellschaftsvertrag* (Wien: Fritz Molden, 1971).


CHAPTER 9. LOVE'S LAWS OF HARMONY

1. Daniel S. Lehrman, "The Reproductive Behavior of Ring Doves," *Scientific American*, Vol. 211, No. 5 (Nov. 1964), pp. 48–54.

2. Report on J. G. Vandenbergh in *Scientific American*, Vol. 226, No. 6 (June 1972), p. 53.

3. Vitus B. Dröscher, *The Magic of the Senses* (New York: E. P. Dutton & Co., Inc., 1969; London: W. H. Allen & Co. Ltd. 1969).

4. Dieter Matthes, "Das Sexualverhalten des Malachiiden Anthocomus coccineus," *Zeitschrift für Tierpsychologie*, 19 (1971), pp. 113–120.

5. Alexander and Margarete Mitscherlich, *Die Unfähigkeit zu trauern—Grundlagen kollektiven Verhaltens* (München: Piper, 1968), p. 193.

6. Ibid.


CHAPTER 10. RULES OF COURTSHIP

1. Wolfgang Wickler, *Sind wir Sünder?* (München: Droemer Knaur, 1969), p. 255.

2. Oskar Heinroth, *Grzimeks Tierleben*, Vol. 9, Vögel 3 (Zürich, 1970), pp. 92–93.

3. Peter Kunkel, "Bemerkungen zu einigen Verhaltenweisen des Rebhuhnastrilds," *Zeitschrift für Tierpsychologie*, Vol. 23 (1966), pp. 136–140.

## CHAPTER 11. GIFTS SEAL FRIENDSHIP

1. Rémy Chauvin, *Tiere unter Tieren* (Bern: Scherz, 1964), p. 214.

2. Friedrich Goethe, *Die Silbermöwe* (Wittenberg: A. Ziemsen, 1956), p. 45.

3. Peter Kunkel, "Bewegungsformen, Sozialverhalten, Balz und Nestbau des Gangesbrillenvogels," *Zeitschrift für Tierpsychologie*, Vol. 19 (1962), pp. 559–576.

## CHAPTER 12. MASS COURTSHIP AND THE MARRIAGE MARKET

1. Vitus B. Dröscher, *The Friendly Beast* (New York. E. P. Dutton & Co., Inc., 1971; London: W. H. Allen & Co., Ltd., 1971).

## CHAPTER 13. WITHOUT POWER THERE IS NO SEXUALITY

1. Richard D. Estes, "Territorial Behavior of the Wildebeest," *Zeitschrift für Tierpsychologie*, Vol. 26 (1969), pp. 284–370.

2. Ibid.

## CHAPTER 14. BEAUTY DISQUALIFIES MALES FROM MARRIAGE

1. The author accompanied his friend Heinrich Hesse.

2. The combs are red or orange patches of skin above the eye. During courtship display they swell up, resembling warts. (Translator's Note.)

3. E. Thomas Gilliard, "The Evolution of Bowerbirds," *Scientific American*, Vol. 209, No. 2 (1963), pp. 38–46.

4. Ibid.

## CHAPTER 15. COURTSHIP AT COURT

1. Robert Watts and Allen W. Stokes, "The Social Order of

Turkeys," *Scientific American,* Vol. 224, No. 6 (1971), pp. 112–118.

2. Ibid.

## CHAPTER 16. THE UNHAPPY LIFE OF A PASHA

1. *Grzimeks Tierleben,* Vol. 12, Säugetiere 3 (Zürich: 1972), pp. 377–378.

2. Vitus B. Dröscher, *The Mysterious Senses of Animals* (New York: E. P. Dutton & Co., Inc., 1964; London: Hodder & Stoughton Ltd., 1964), pp. 80–90.

3. Hans Kummer, "Ursachen von Gesellschaftsformen bei Primaten," *Umschau in Wissenschaft und Technik,* 1972, pp. 481–484. Hans Kummer, "Immediate causes of primate social structure," Proceedings of the Third International Congress on Primates, 3, pp. 1–11. Basel, Karger, 1971.

4. A personal communication from Adriaan Kortlandt to the author.

## CHAPTER 17. MARRIAGE PARTNERS WHO DO NOT KNOW EACH OTHER

1. The author's own observations.

2. Juan D. Delius, "Das Verhalten der Feldlerche," *Zeitschrift für Tierpsychologie,* Vol. 20 (1963), pp. 297–348.

3. Ibid.

## CHAPTER 18. ONLY AGGRESSIVE MATES STAY TOGETHER

1. Rudolf Berndt and Helmut Sternberg, "Paarbildung und Partneralter beim Trauerschnäpper," *Die Vogelwarte,* Vol. 26 (1971), pp. 136–142.

2. Ibid.

## CHAPTER 19. IS FIDELITY AN ILLUSION?

1. R. A. Stamm, "Aspekte des Paarverhaltens von Agapornis personata," *Behaviour,* 19, pp. 1–56.

2. Ibid.

3. Otto Koenig, "Das Aktionssystem der Bartmeise," *Oster-reichische Zoologische Zeitschrift,* 3 (1954).

4. Desmond Morris, *The Naked Ape* (New York: McGraw-Hill Publishing Co., Inc., 1967; London: Jonathan Cape, 1967).

5. See Note 3.

## CHAPTER 20. FROM ANIMAL TO MAN

1. Vitus B. Dröscher, *The Friendly Beast* (New York: E. P. Dutton & Co., Inc., 1971; London: W. H. Allen & Co., Ltd., 1971).

2. Jürg Lamprecht, "Duettgesang beim Siamang," *Zeitschrift für Tierpsychologie,* 27 (1970), pp. 186–204.

3. Ibid.

## CHAPTER 21. CAN BROTHERS MARRY SISTERS?

1. *Scientific American,* Vol. 227, No. 6 (1972), p. 43.

## CHAPTER 22. ADULTERY AND DIVORCE

1. J. C. Coulson, "The Influence of the Pair-Bond and Age on the Breeding Biology of the Kittiwake Gull." *Journal of Animal Ecology,* 35, pp. 269–279.

2. Ibid.

3. Hubert Markl, "Vom Eigennutz des Uneigennützigen," *Naturwissenschaftliche Rundschau,* 24 (1971), pp. 281–289.

# Index